MONDAY Patterning and

1. Draw the next shape in the pattern.
   Colour the shapes to make another pattern.

______

2. There were 3 bugs.
   2 more bugs came.
   How many bugs are
   there now?

3 plus 2 ______ in all

TUESDAY Number Sense and Operations

1. How many tens and ones?

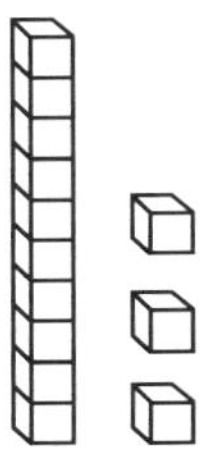

tens ______

ones ______

2. Write the numeral.

four ______

3. How many?

______

4. Circle the set that has more.

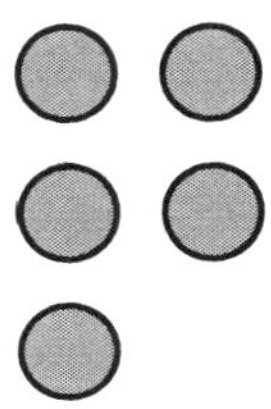

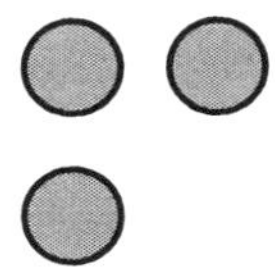

WEDNESDAY Geometry

1. Circle the name of this shape.

   circle   triangle

2. How many sides does it have? ________

3. How many vertices does it have? ________

4. Trace and then draw the shape.

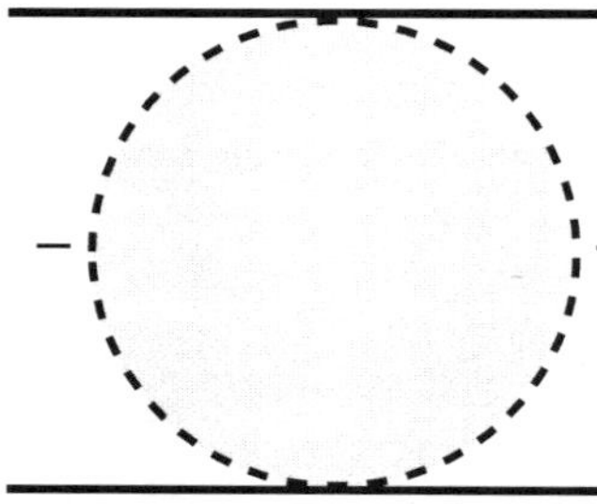

THURSDAY Measurement

1. What time is it?

________ o'clock

2. Circle the container that holds more.

3. Which flower is the biggest?

A.

B.

C.

FRIDAY Data Management

Favourite Pets

1. How many? 

2. Circle the most popular pet. 

3. Circle the least popular pet. 

BRAIN STRETCH

1. 

______ + ______ = ______

2. 

______ − ______ = ______

MONDAY Patterning and Algebra

1. Draw the next shape in the pattern.
   Colour the shapes to make another pattern.

2. There were 5 stamps.
   Ellie used 3 stamps.
   How many are left?

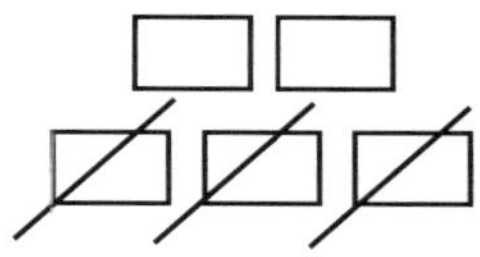

5 take away 3 ______ are left

TUESDAY Number Sense and Operations

1. How many tens and ones?

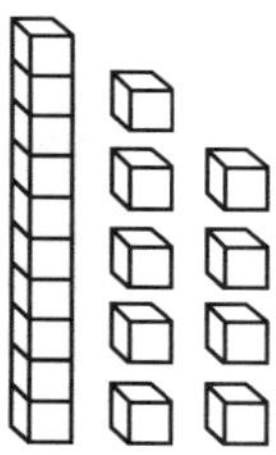

tens ______

ones ______

2. What is the value of a nickel?

______ ¢

3. Write the numeral.

one ______

4. Circle the set that has fewer.

WEDNESDAY Geometry

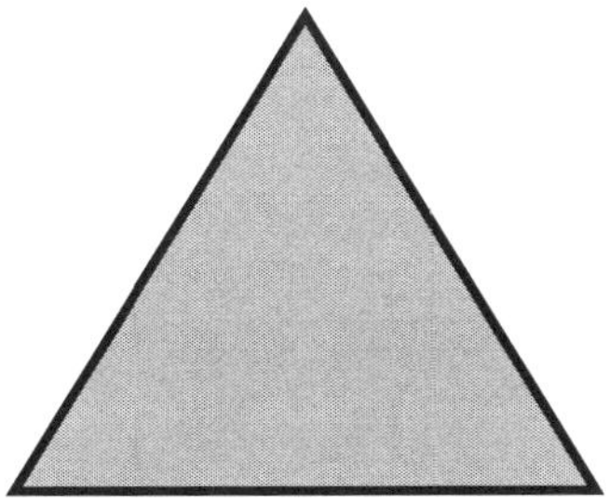

1. Circle the name of this shape.

   circle triangle

2. How many sides does it have? ________

3. How many vertices does it have? ________

4. Trace and then draw the shape.

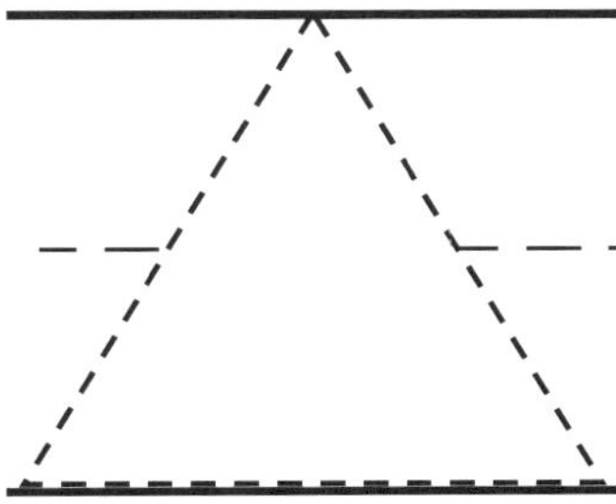

THURSDAY Measurement

1. What time is it?

________ o'clock

2. Which tree is shorter?

A. 

B. 

3. Which  is the tallest?

A. 

B. 

C. 

FRIDAY Data Management

1. Complete the tally chart.

Favourite Fish

	Tally	Number
	𝍸 I	
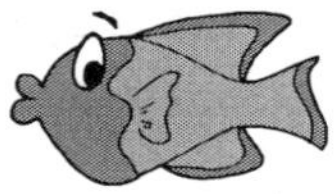	𝍸 𝍸	
	IIII	

2. Circle the most popular fish.

3. Circle the least popular fish.

BRAIN STRETCH

1.

_______ + _______ = _______

2.

_______ – _______ = _______

MONDAY Patterning and Algebra

1. Draw the next shape in the pattern.
   Colour the shapes to make another pattern.

______

2. Sam has 1 marble.
   He got 6 more marbles.
   How many marbles
   does Sam have now?

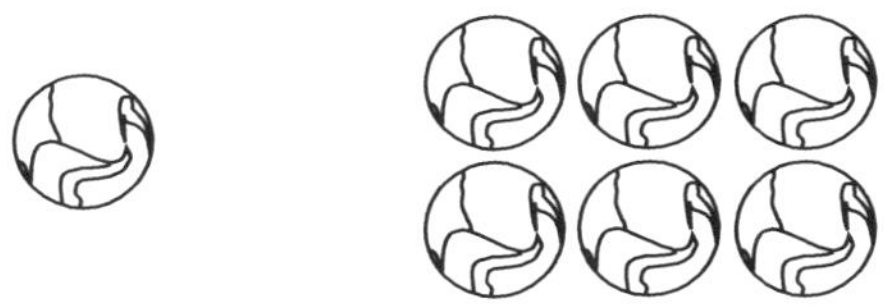

1 plus 6 ______ in all

TUESDAY Number Sense and Operations

1. How many tens and ones?

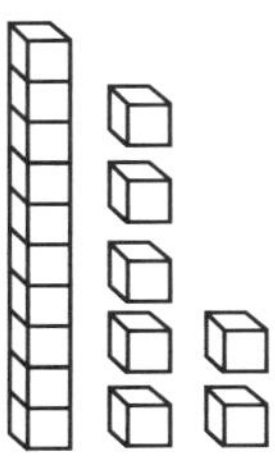

tens ______

ones ______

2. What is the value of a dime?

______ ¢

3. How many?

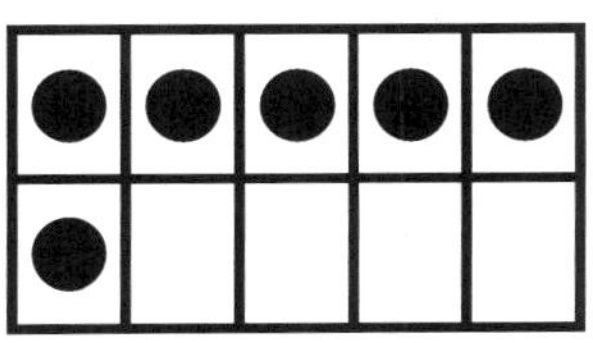

______

4. Circle the set that has more.

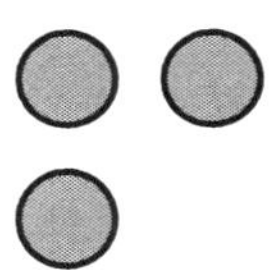

WEDNESDAY Geometry

1. Circle the name of this shape.

   circle   square

2. How many sides does it have? ________

3. How many vertices does it have? ________

4. Trace and then draw the shape.

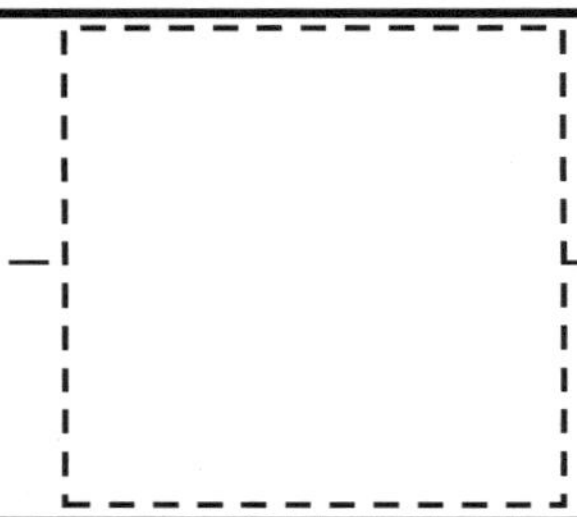

THURSDAY Measurement

1. What time is it?

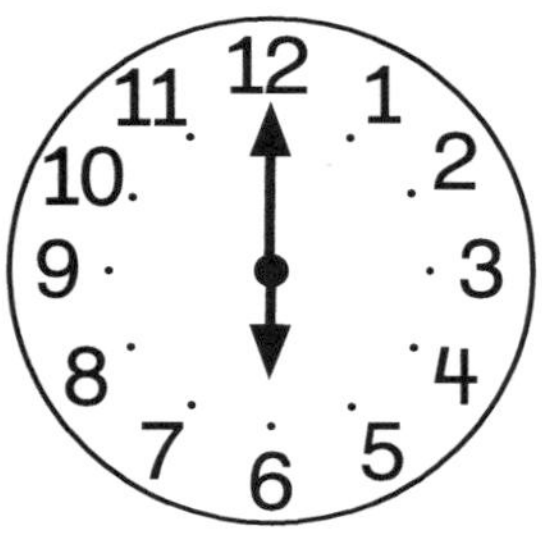

_____ : _____

2. Circle the container that holds less.

A. 

B. 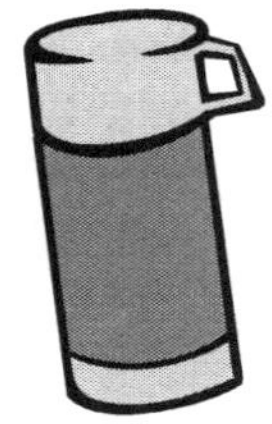

3. Measure the length of the line.

It is about _________  long.

FRIDAY Data Management

Favourite Toy

1. How many? 

2. Circle the most popular toy. 

3. Circle the least popular toy. 

BRAIN STRETCH

1. 

_______ + _______ = _______

2. 

_______ – _______ = _______

MONDAY Patterning and Algebra

1. Draw the next shape in the pattern.
   Colour the shapes to make another pattern.

______

2. Lisa has 1 orange
   and Pat has 8 oranges.
   How many oranges
   are there in all?

1 plus 8 ______ in all

TUESDAY Number Sense and Operations

1. How many tens and ones?

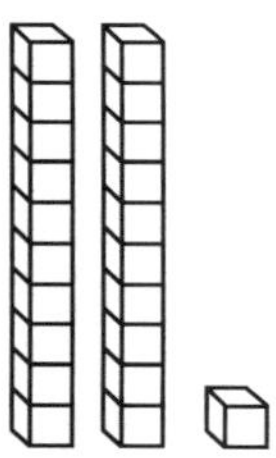

tens ______

ones ______

2. What is the value of a quarter?

______ ¢

3. What is the number after 3?

2, 3, ______

4. Circle the set that has fewer.

WEDNESDAY Geometry

1. Circle the name of this shape.

   rectangle   triangle

2. How many sides does it have? _________

3. How many vertices does it have? _________

4. Trace and then draw the shape.

THURSDAY Measurement

1. What time is it?

_____ : _____

2. Which animal is bigger?

A.    B. 

3. Measure the length of the line.

It is about _______  long.

FRIDAY Data Management

1. Complete the tally chart.

Favourite Playground Equipment

	Tally	Number
	𝍸	
	𝍸 ||	
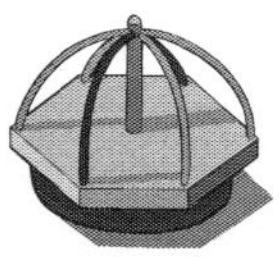	𝍸 |	

2. Circle the most popular playground equipment.

  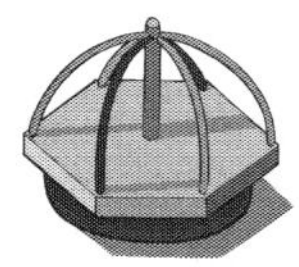

3. Circle the least popular playground equipment.

  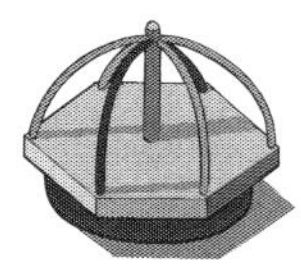

BRAIN STRETCH

1.

_______ + _______ = _______

2.

_______ − _______ = _______

MONDAY Patterning and Algebra

1. Colour the shapes to make your own pattern.
   Circle the core of the pattern.

☐ ☐ ☐ ☐ ☐ ☐

2. There were 4 frogs.
   Then 2 frogs left.
   How many are left?

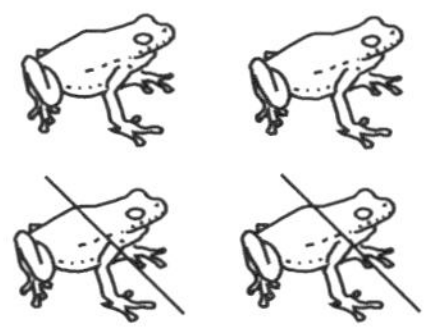

4 take away 2 ______ are left

TUESDAY Number Sense and Operations

1. How many tens and ones?

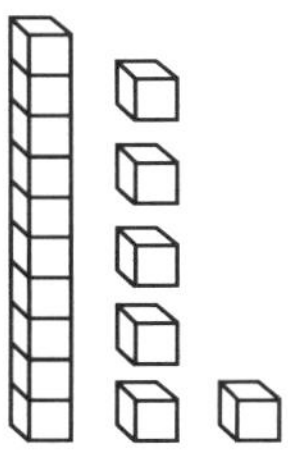

tens ______

ones ______

2. What is the value of the coin?

______ ¢

3. Write the numeral.

two ______

4. Circle the set that has more.

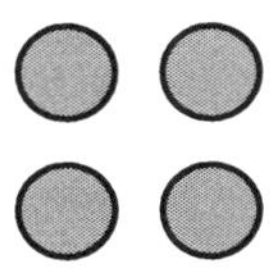

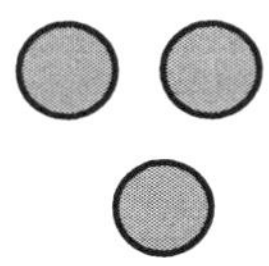

WEDNESDAY Geometry

1. Circle the name of this shape.

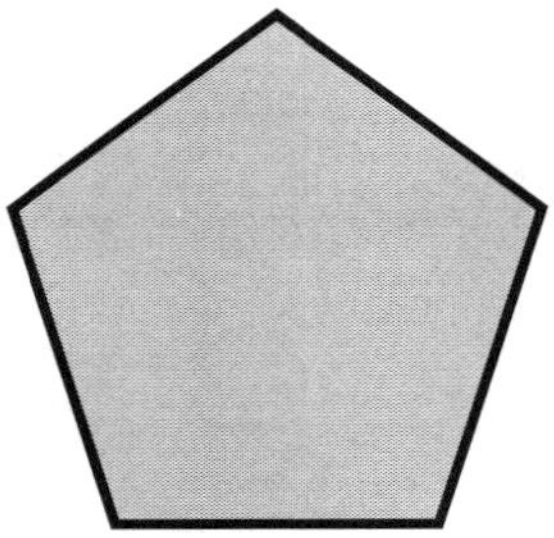

pentagon   rectangle

2. How many sides does it have? ________

3. How many vertices does it have? ________

4. Trace and then draw the shape.

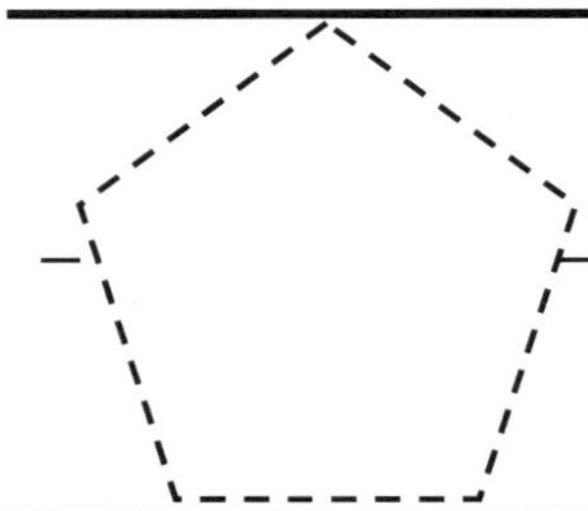

THURSDAY Measurement

1. What time is it?

________ o'clock

2. Which tree is the tallest?

A.   B.   C.

3. Measure the length of the line.

It is about ________  long.

FRIDAY Data Management

Favourite Fruit

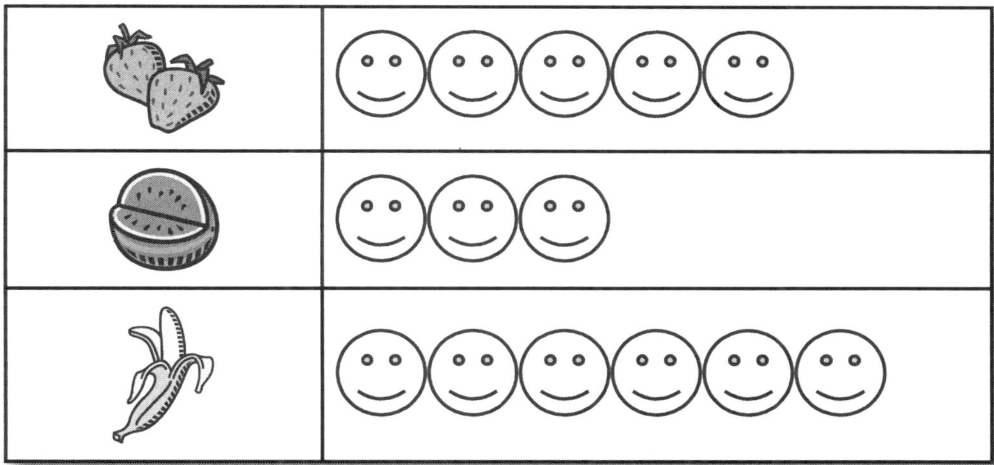

1. How many? 

2. Circle the most popular fruit. 

3. Circle the least popular fruit. 

BRAIN STRETCH

Colour one half of each shape.

A. 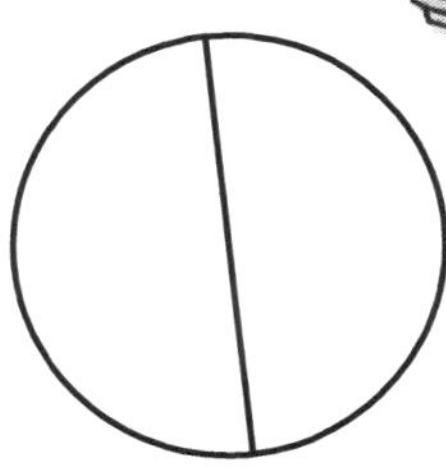

B. 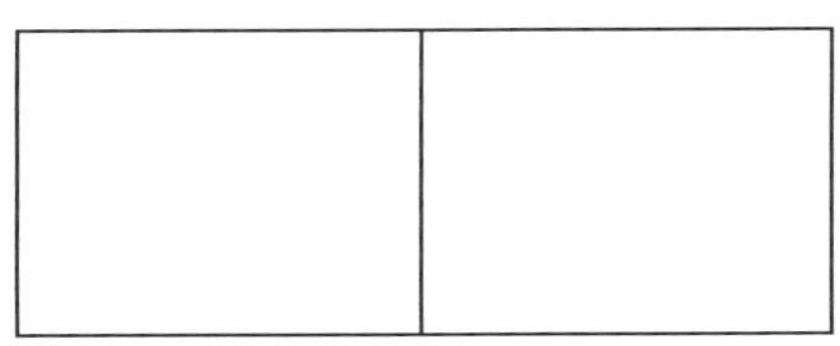

MONDAY Patterning and Algebra

1. Colour the shapes to make a pattern. Circle the core of the pattern.

2. There were 6 birds in the tree. Then 3 birds flew away. How many are left?

6 take away 3 _____ are left

TUESDAY Number Sense and Operations

1. How many tens and ones?

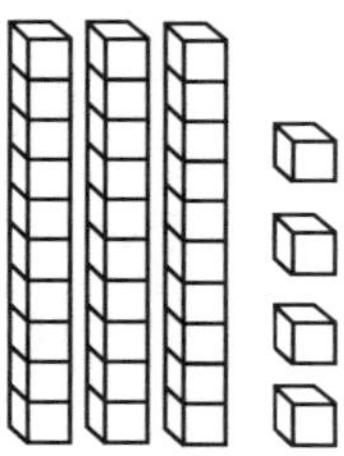

tens ______

ones _______

2. What is the value of the coin?

_________ ¢

3. Write the numeral.

ten _________

4. Circle the smaller number.

6 3

WEDNESDAY Geometry

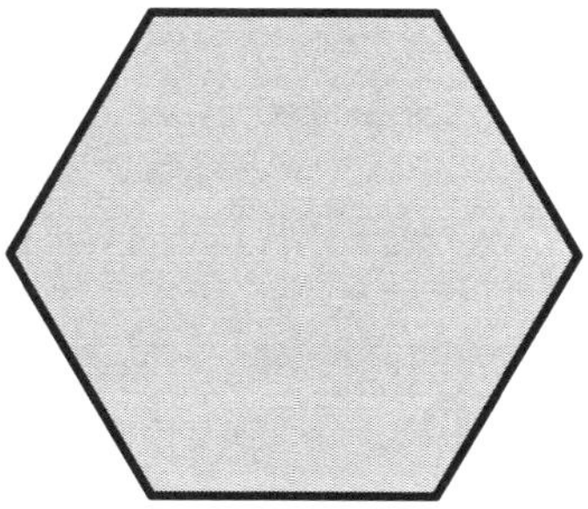

1. Circle the name of this shape.

   hexagon   rectangle

2. How many sides does it have? ________

3. How many vertices does it have? ________

4. Trace and then draw the shape.

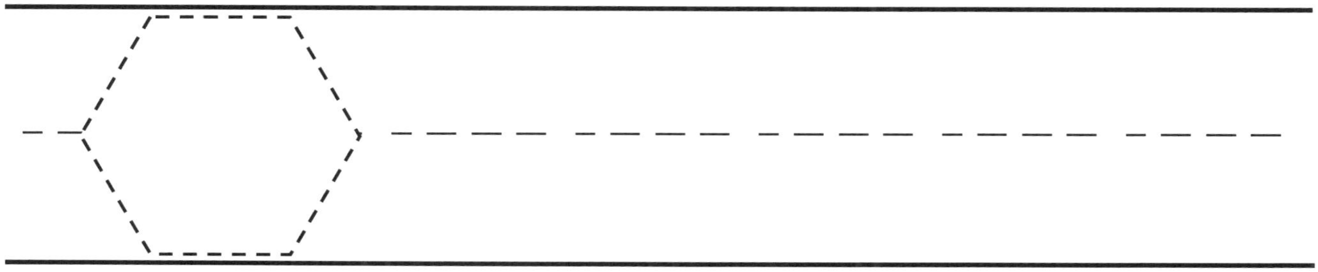

THURSDAY Measurement

1. What time is it?

_____ : _____

2. Is the temperature hot or cold?

A. hot   B. cold

3. Measure the length of the line.

It is about ________  long.

FRIDAY Data Management

1. Complete the tally chart.

Favourite Vegetable

	Tally	Number
	𝍸 I	
	II	
	𝍸 III	

2. Circle the most popular vegetable.   

3. Circle the least popular vegetable.   

BRAIN STRETCH

Count from 1 to 10.

1				
				10

MONDAY Patterning and Algebra

1. Make a pattern using  and . Circle the core of the pattern.

____

2. Colour the blocks using two colours to show ways to make 5.
Complete each addition sentence.

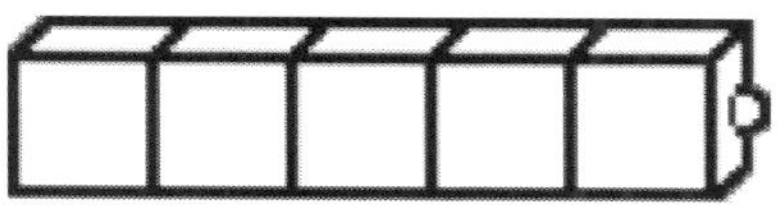

_______ + _______ = 5

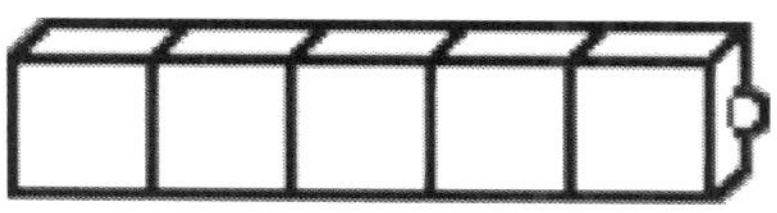

_______ + _______ = 5

TUESDAY Number Sense and Operations

1. How many tens and ones?

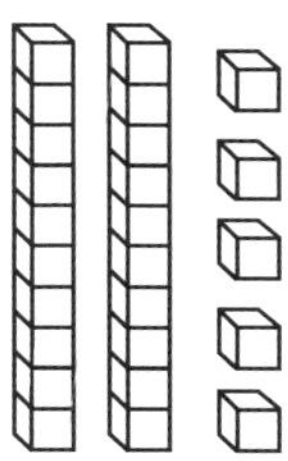

tens _____

ones ______

2. What is the value of the coins?

________ ¢

3. What is the number before 8?

______, 8, 9

4. Circle the set that has more.

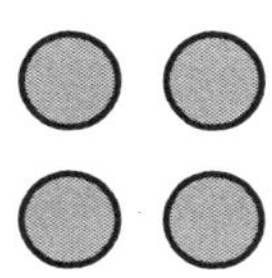

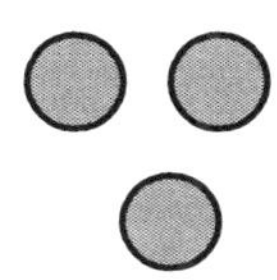

WEDNESDAY Geometry

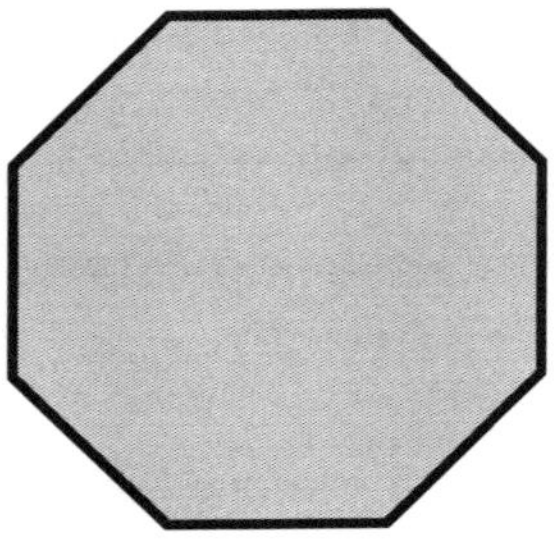

1. Circle the name of this shape.

   octagon triangle

2. How many sides does it have? ________

3. How many vertices does it have? ________

Trace and then draw the shape.

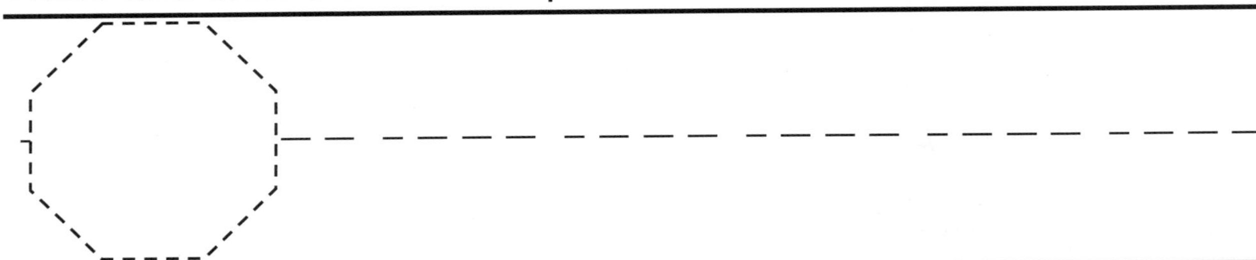

THURSDAY Measurement

1. What time is it?

______ o'clock

2. Is the temperature hot or cold?

A. hot B. cold

3. Measure the length of the line.

It is about ________  long.

FRIDAY Data Management

Favourite Treats

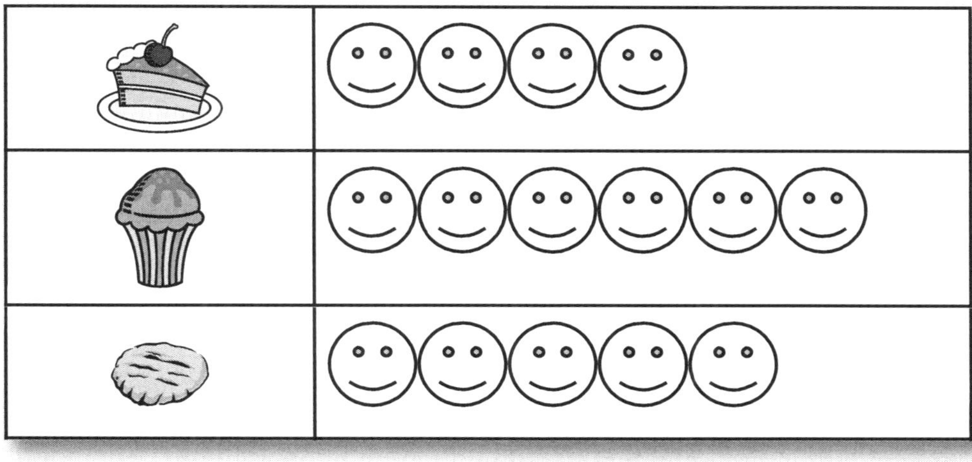

1. How many?  ________  ________  ________

2. Circle the most popular treat.   

3. Circle the least popular treat.   

BRAIN STRETCH

What is the number?

A. 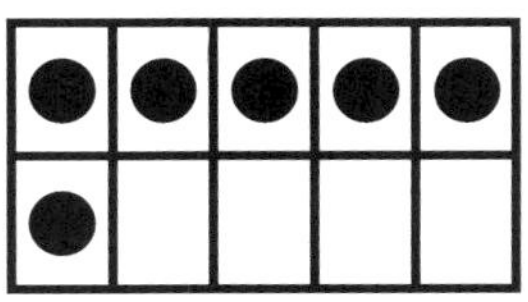 ________

B. 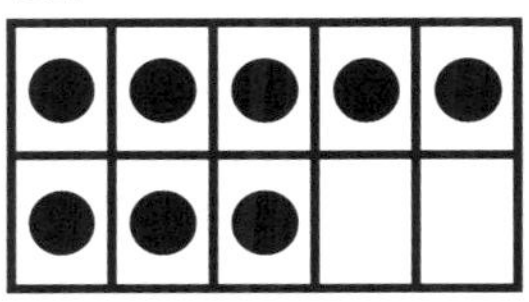 ________

C. 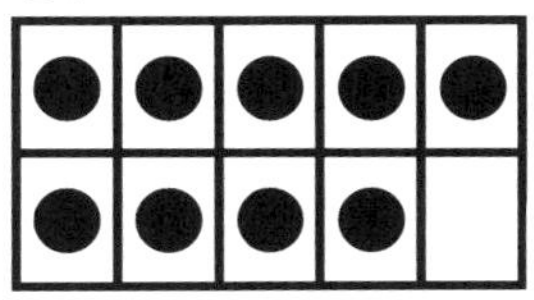 ________

MONDAY Patterning and Algebra

1. Make a pattern. Circle the core of the pattern.

____

2. Colour the blocks using two colours to show ways to make 6. Complete each addition sentence.

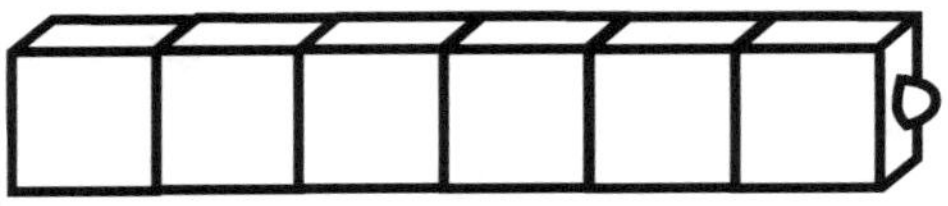

_______ + _______ = 6

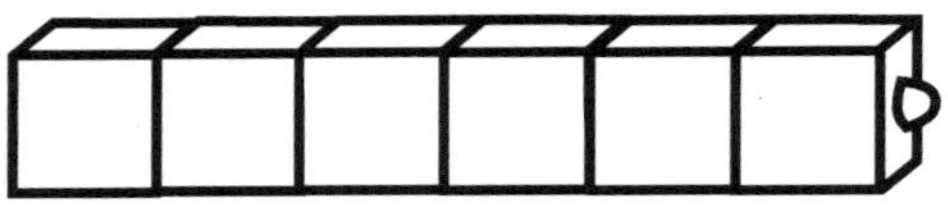

_______ + _______ = 6

TUESDAY Number Sense and Operations

1. How many tens and ones?

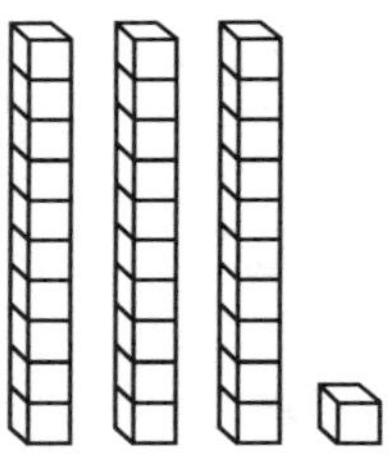

tens _____

ones _____

2. What is the value of the coins?

_______ ¢

3. How many?

____

4. What is the number between 2 and 4?

2, _____, 4

WEDNESDAY Geometry

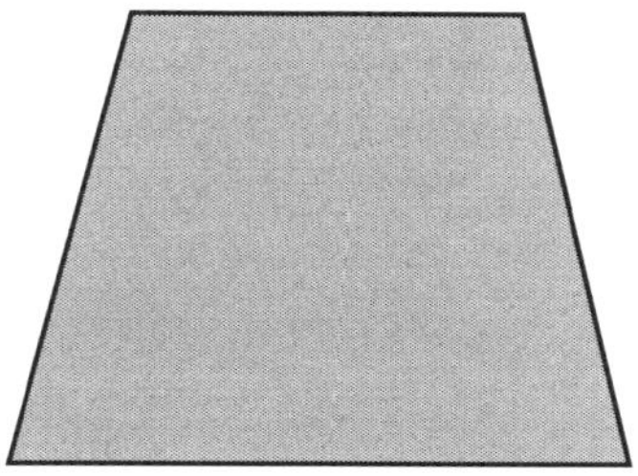

1. Circle the name of this shape.

   circle   trapezoid

2. How many sides does it have? __________

3. How many vertices does it have? __________

4. Trace and then draw the shape.

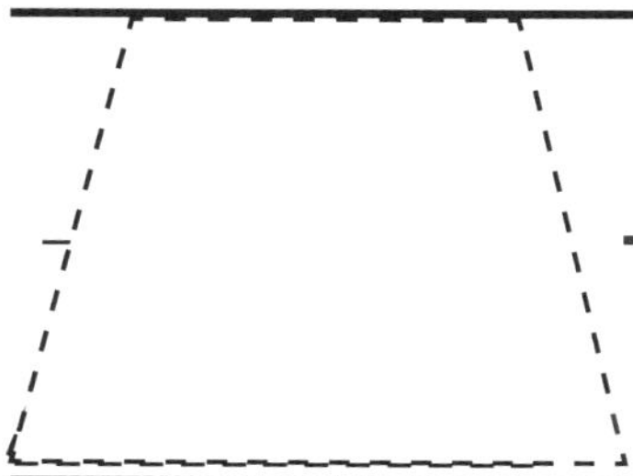

THURSDAY Measurement

1. What time is it?

______ : ______

2. How many days in a week?

____

3. Measure the length of the line.

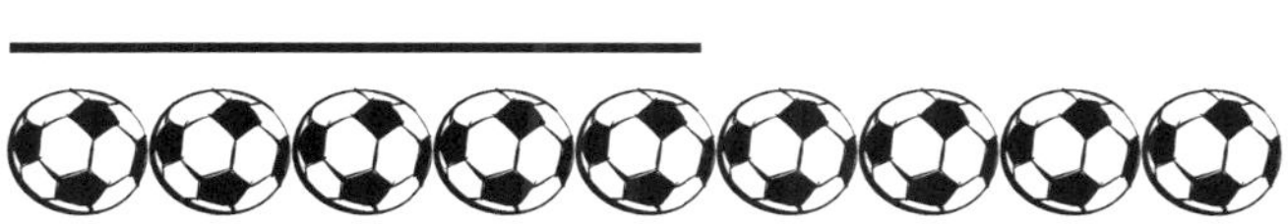

It is about ________  long.

FRIDAY Data Management

1. Complete the tally chart.

Favourite Weather

	Tally	Number
	𝍸 ||||	
	|||	
	𝍸 |	

2. Circle the most popular kind of weather.

3. Circle the least popular kind of weather.

BRAIN STRETCH

Write true or false.
Draw a picture to check.

A. 6 = 5 – 1 _____

B. 6 = 7 – 1 _____

MONDAY Patterning and Algebra

1. Make an AB pattern. Circle the core of the pattern.

____

2. Cross out the blocks you want to take away.
   Colour the blocks left. Complete the subtraction sentence.

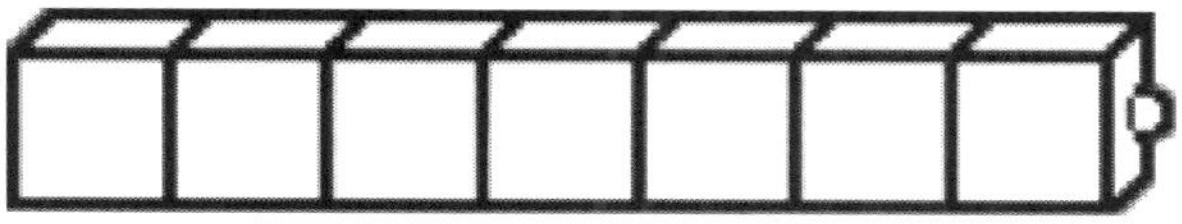

7 − ________ = ________

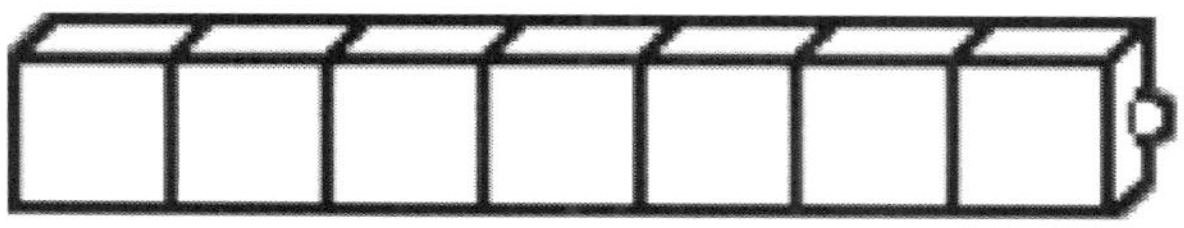

7 − ________ = ________

TUESDAY Number Sense and Operations

1. How many tens and ones?

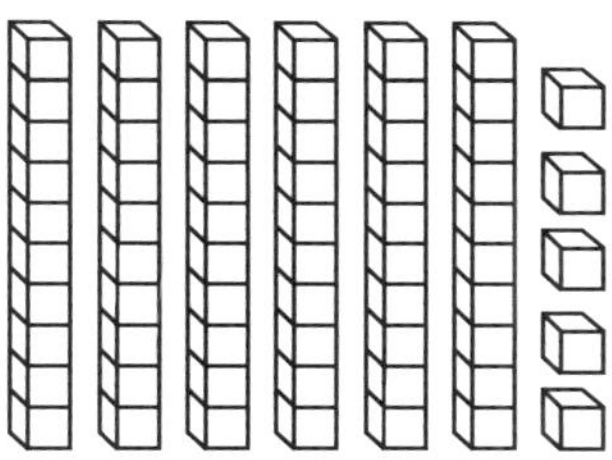

tens ______

ones ______

2. What is the value of the coins?

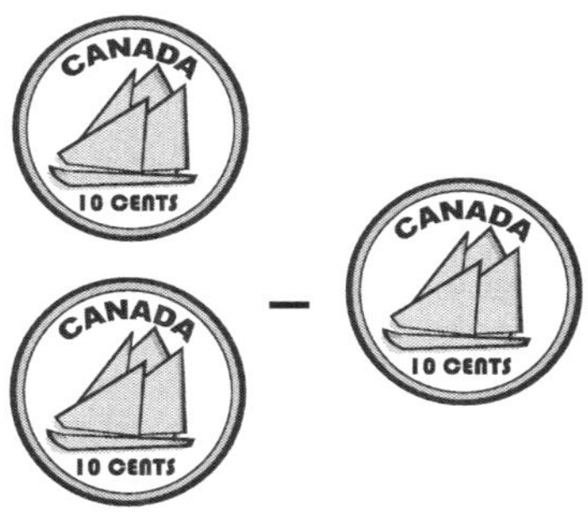

= ________ ¢

3. What is the number after 7?

6, 7, _____

4. How many?

____

WEDNESDAY Geometry

Colour the square green.
Colour the triangle yellow.

Colour the circle blue.
Colour the rectangle red.

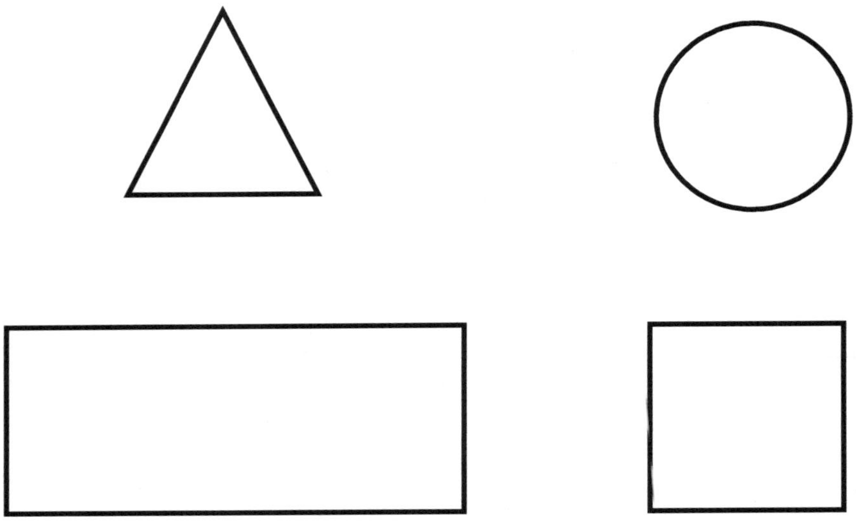

THURSDAY Measurement

1. What time is it?

______ : ______

2. What day of the week comes after Monday?

A. Tuesday B. Friday

3. Measure the length of the line.

It is about ________  long.

FRIDAY Data Management

Favourite Footwear

1. How many? 

2. Circle the most popular footwear. 

3. Circle the least popular footwear. 

BRAIN STRETCH

There were 3 kittens in the garden.
At the end, there were 9 kittens in all.
How many kittens came?

Draw a picture.

3 + ___ = 9

So, ___ kittens came.

MONDAY Patterning and Algebra

1. Make an AABB pattern using  and . Circle the core of the pattern.

____

2. Cross out the blocks you want to take away.
   Colour the blocks left. Complete the subtraction sentence.

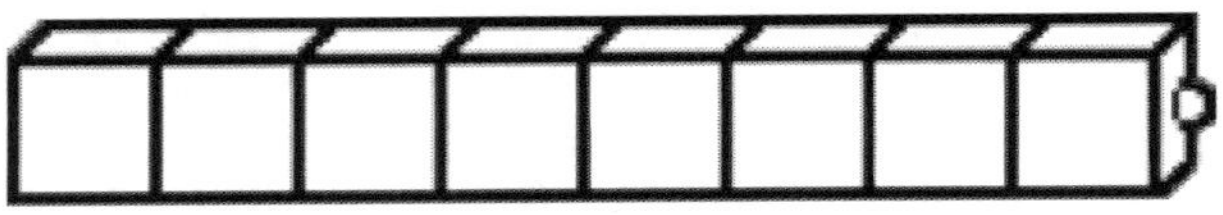

8 − ________ = ________

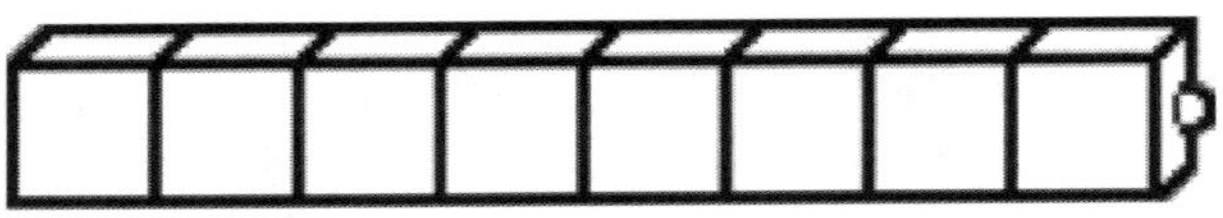

8 − ________ = ________

TUESDAY Number Sense and Operations

1. How many tens and ones?

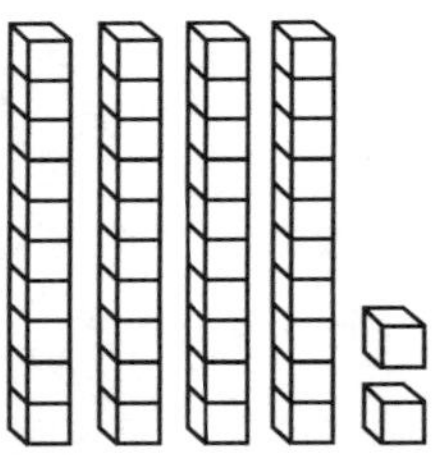

tens ______

ones ______

2. What is the value of the coins?

________ ¢

3. What is the number between 8 and 10?

   8, ______, 10

4. Draw a 10 stick. The number 10 has ____ ten and 0 ones.

WEDNESDAY Geometry

How many? ○ ______ △ ______ □ ______

THURSDAY Measurement

1. What time is it?

______ o'clock

2. What day of the week comes after Saturday?

A. Sunday B. Wednesday

3. Measure the length of the line.

It is about ________ ⚽ long.

FRIDAY Data Management

1. Complete the tally chart.

Favourite Activity

	Tally	Number
	卌 I	
	卌 II	
	卌 IIII	

2. Circle the least popular activity.

BRAIN STRETCH

Use the number line to count on to find each sum.

1. $\begin{array}{r} 6 \\ +\ 3 \\ \hline \end{array}$

2. $\begin{array}{r} 1 \\ +\ 2 \\ \hline \end{array}$

3. $\begin{array}{r} 2 \\ +\ 1 \\ \hline \end{array}$

4. $\begin{array}{r} 7 \\ +\ 5 \\ \hline \end{array}$

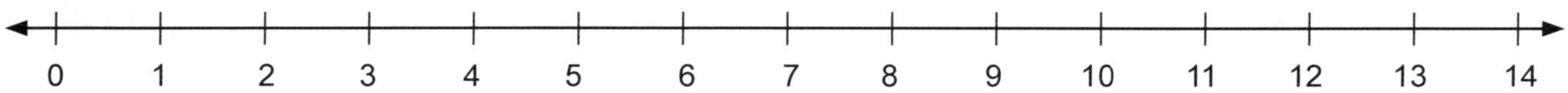

MONDAY Patterning and Algebra

1. Show three ways to make 4. Colour the blocks using two colours.

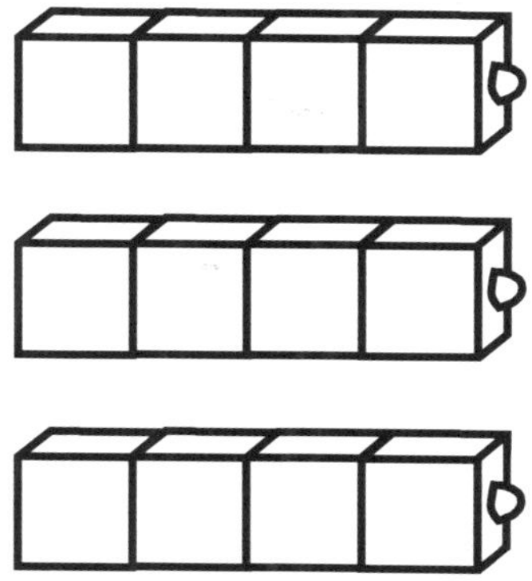

___ + ___ = 4

___ + ___ = 4

___ + ___ = 4

2. There were 7 blocks. Dan took some blocks. Then there were 3 blocks left.

   Cross out the blocks left.
   Dan took ___ blocks.

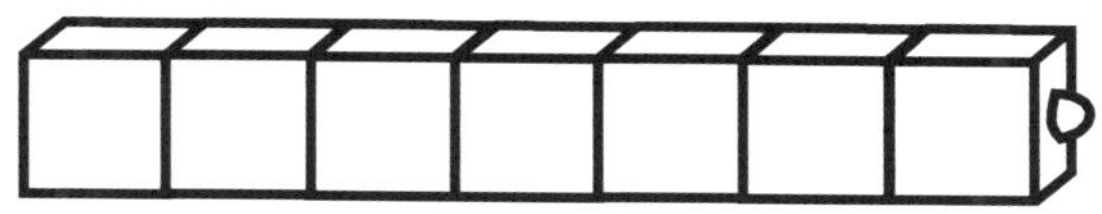

TUESDAY Number Sense and Operations

1. How many tens and ones?

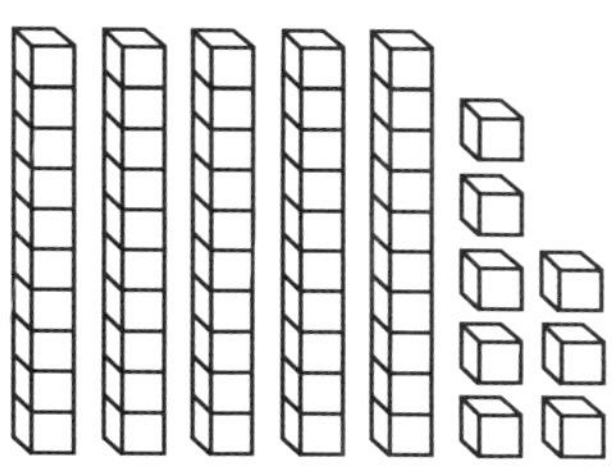

tens ______

ones ______

2. There are 13 cubes. Draw a circle around a group of 10 cubes. How many cubes are left?

   ___ cubes

   So, the number 13 has ___ ten and ___ ones.

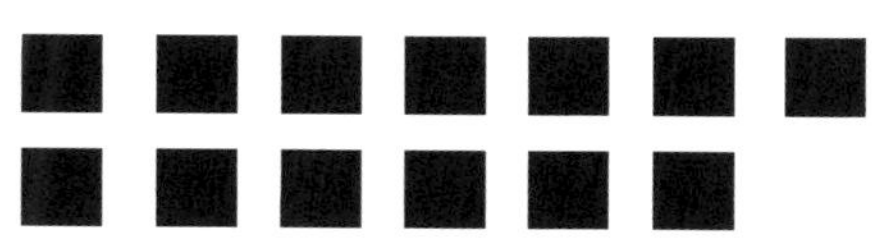

3. What is the number between 12 and 14?

   12, ______, 14

WEDNESDAY Geometry

Read and follow the sorting rule.

1. Colour the shapes with more than 3 sides.

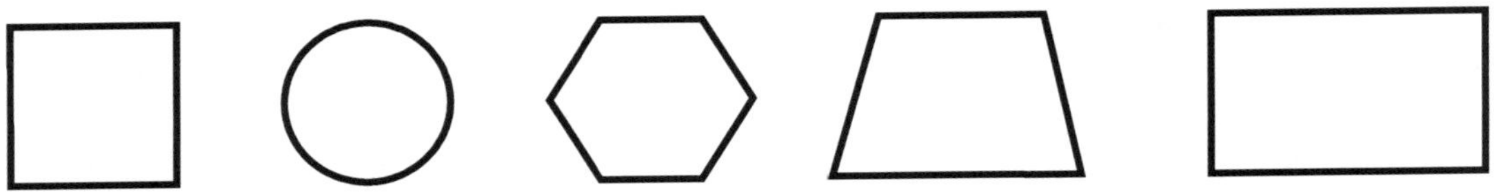

2. Colour the shapes with 4 vertices.

THURSDAY Measurement

1. What time is it?

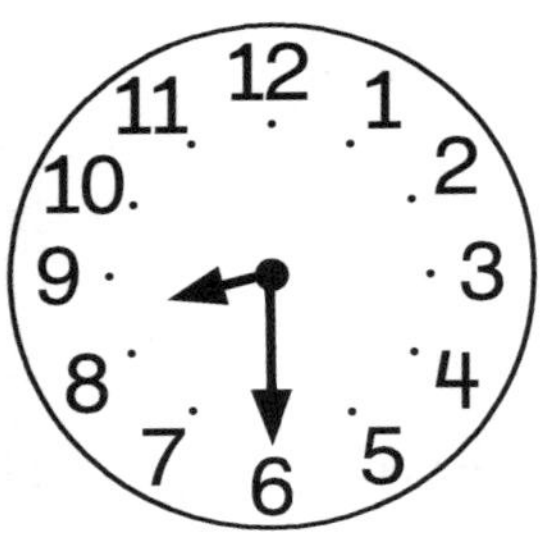

half past ______

2. What day of the week comes before Wednesday?

A. Monday B. Tuesday

3. Measure the length of the line.

It is about ________  long.

FRIDAY Data Management

Favourite Shape

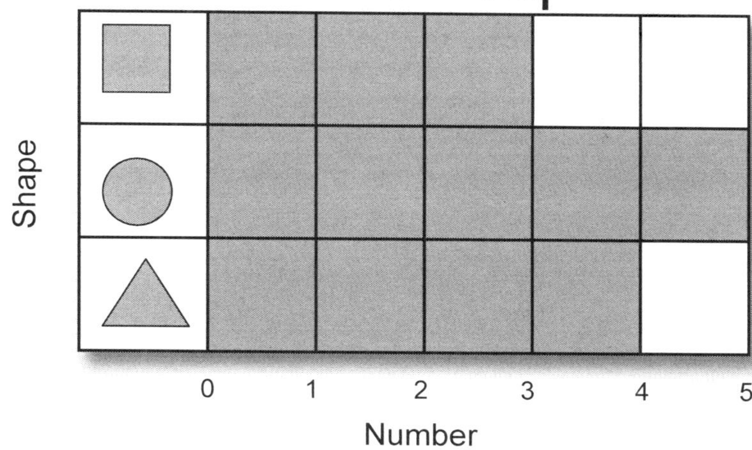

1. How many? ___ ___ ___

2. Circle the most popular shape.

BRAIN STRETCH

Use the number line to count back to find each difference.

1. $\begin{array}{r} 4 \\ -\ 3 \\ \hline \end{array}$

2. $\begin{array}{r} 8 \\ -\ 2 \\ \hline \end{array}$

3. $\begin{array}{r} 6 \\ -\ 4 \\ \hline \end{array}$

4. $\begin{array}{r} 9 \\ -\ 3 \\ \hline \end{array}$

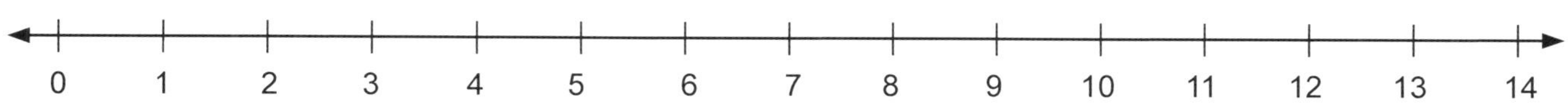

MONDAY Patterning and Algebra

1. Colour an ABC pattern. Circle the core of the pattern.

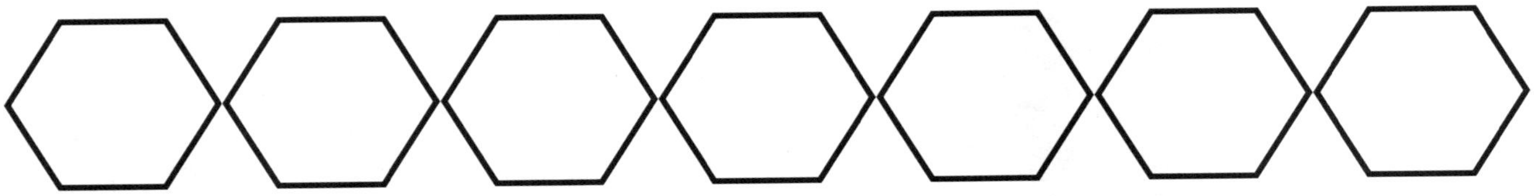

2. Make a pattern using  and . Circle the core of the pattern.

____

3. Describe your pattern.

____

TUESDAY Number Sense and Operations

1. How many tens and ones?

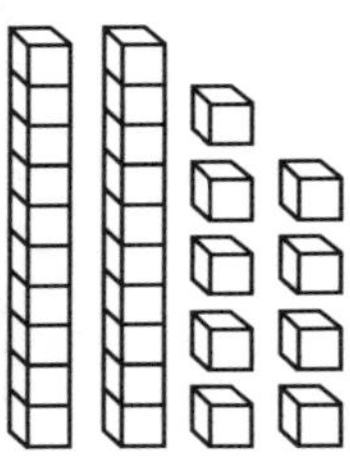

tens ______

ones ______

2. Draw coins to make 20 cents.

3. Circle the third monster.

4. How many?

______

WEDNESDAY Geometry

1. Colour the hexagon green.
   Colour the circle yellow.
   Colour the triangle blue.
   Colour the trapezoid red.

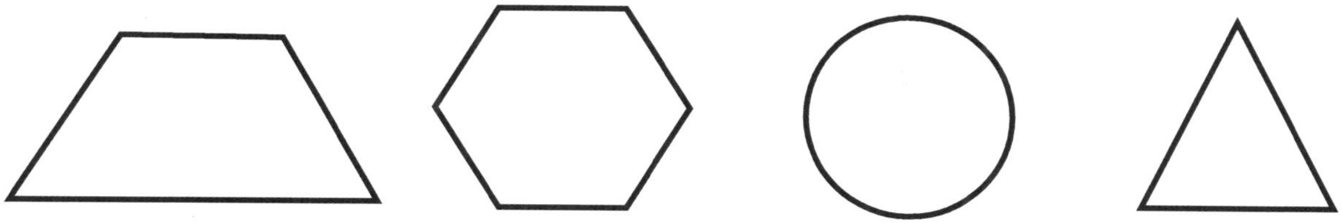

2. What 3D shape can you make using squares?

A. 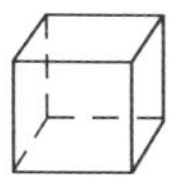

B. 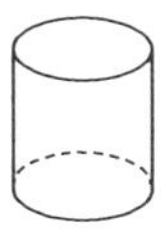

3. What new shape can you make using these 2 shapes?

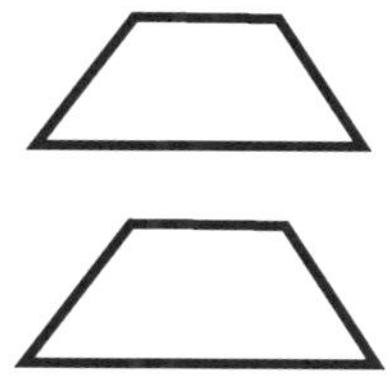

A. 

B. 

THURSDAY Measurement

1. What time is it?

half past _____

2. What day of the week comes after Wednesday?

A. Thursday B. Monday

3. Measure the length of the line.

It is about _________  long.

FRIDAY Data Management

Favourite Sport

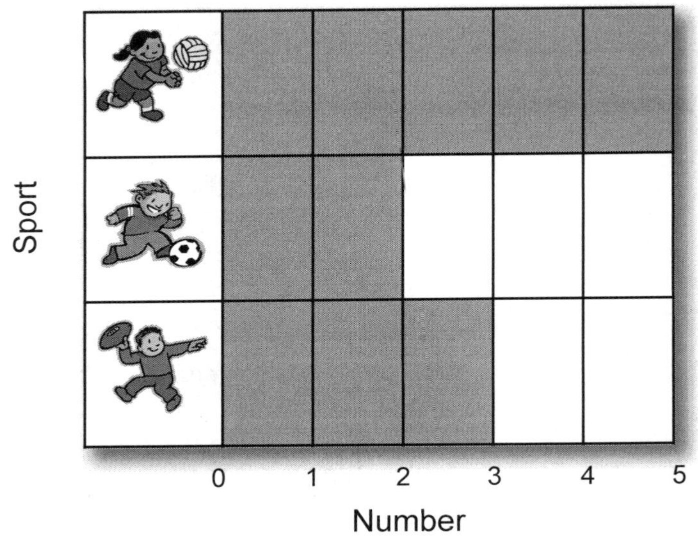

1. How many? 

2. Circle the least popular sport. 

BRAIN STRETCH

Use the number line to count on to find each sum.

1.
$$\begin{array}{r} 4 \\ +\ 4 \\ \hline \end{array}$$

2.
$$\begin{array}{r} 8 \\ +\ 6 \\ \hline \end{array}$$

3.
$$\begin{array}{r} 9 \\ +\ 7 \\ \hline \end{array}$$

4.
$$\begin{array}{r} 7 \\ +\ 2 \\ \hline \end{array}$$

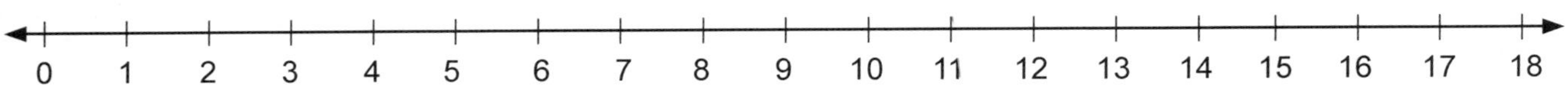

MONDAY Patterning and Algebra

1. Colour an AAB pattern. Circle the core of the pattern.

2. Show three ways to make 9. Colour the blocks using two colours.

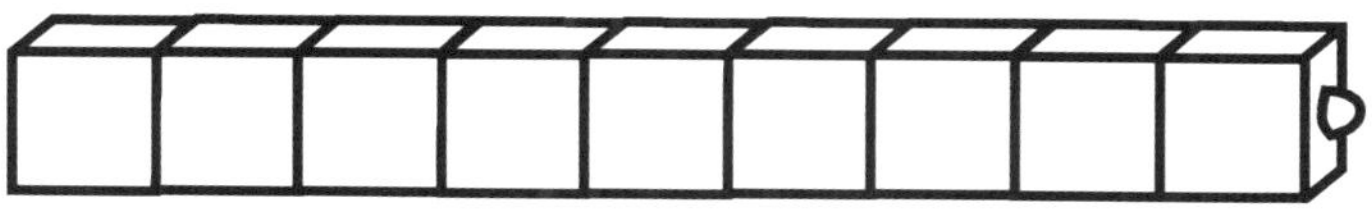

___ + ___ = 9

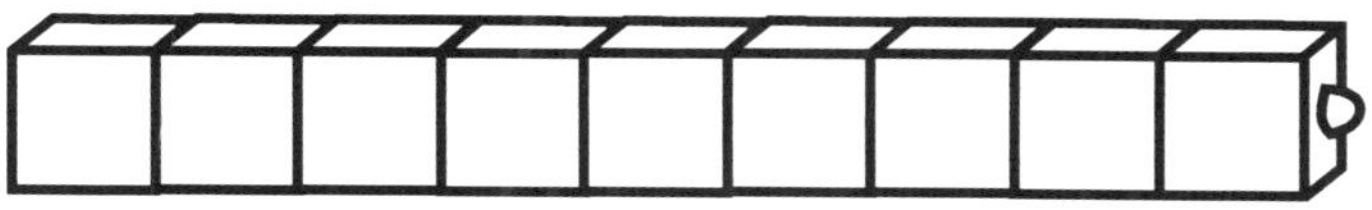

___ + ___ = 9

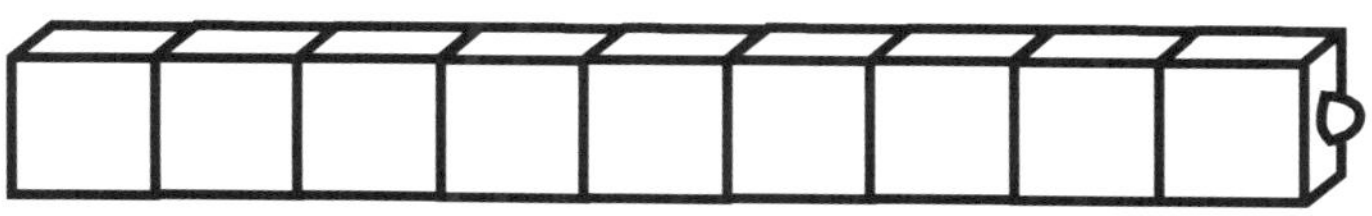

___ + ___ = 9

TUESDAY Number Sense and Operations

1. How many tens and ones?

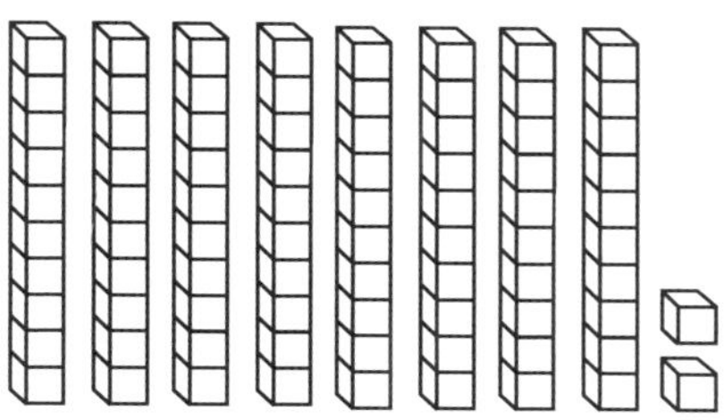

tens ______

ones ______

2. Draw coins to make 15 cents.

3. Circle true or false.

42 is greater than 30

A. True B. False

4. Circle true or false.
The number 40 is made up of 4 tens and 0 ones.

A. True B. False

WEDNESDAY Geometry

1. Draw 2 shapes that have 4 sides.

2. What new shape can you make using these 2 shapes?

A. 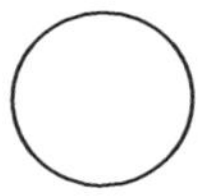

B. 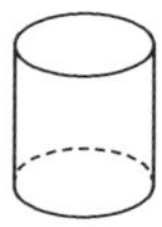

3. What new shape can you make using these 2 shapes?

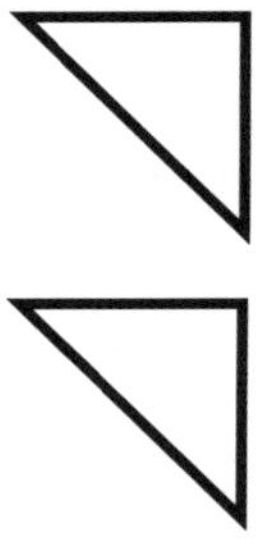

A. 

B. 

THURSDAY Measurement

1. What time is it?

half past _____

2. What day of the week comes after Sunday?

A. Monday B. Tuesday

3. Measure the length of the line.

It is about _________  long.

FRIDAY Data Management

Favourite Winter Activity

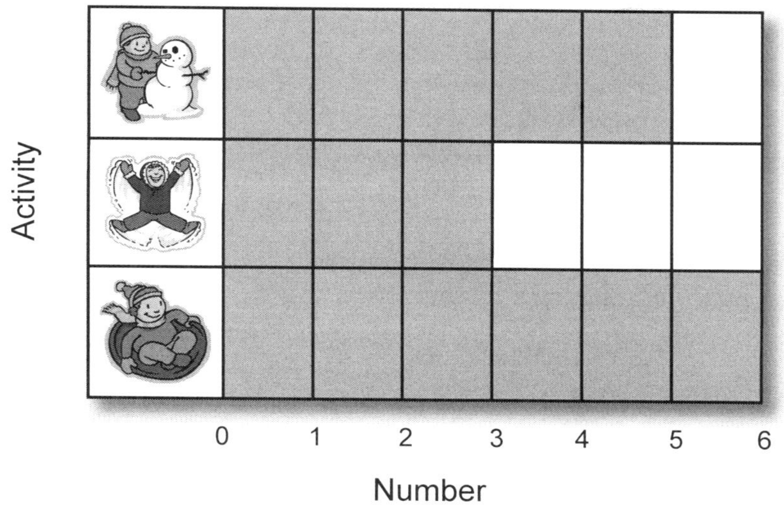

1. Circle the most popular winter activity. 

2. Circle the least popular winter activity. 

BRAIN STRETCH

Use the number line to count back to find each difference.

1. $\begin{array}{r} 16 \\ -\ 8 \\ \hline \end{array}$

2. $\begin{array}{r} 13 \\ -\ 4 \\ \hline \end{array}$

3. $\begin{array}{r} 10 \\ -\ 5 \\ \hline \end{array}$

4. $\begin{array}{r} 18 \\ -\ 9 \\ \hline \end{array}$

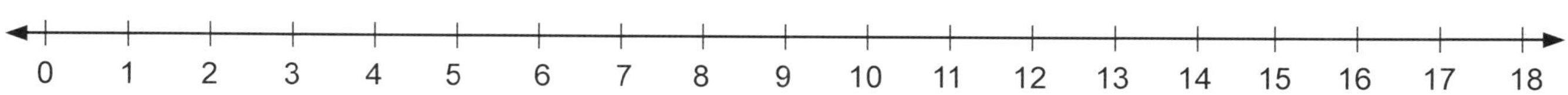

MONDAY Patterning and Algebra

1. Colour an AABB pattern. Circle the core of the pattern.

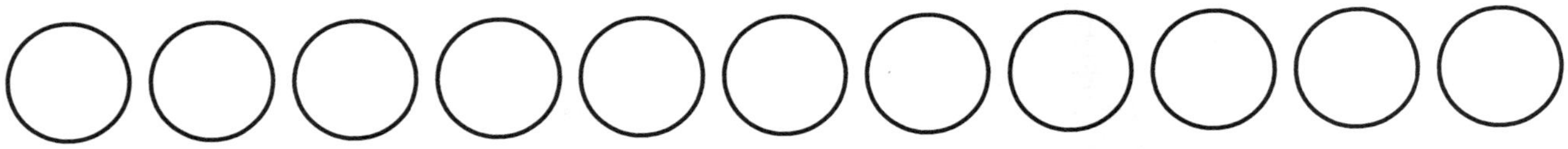

2. Amy has 7 stickers. Brad has 2 more stickers than Amy.
   How many stickers does Brad have?
   Draw 7 counters. Then draw 2 more counters.

   7 + 2 = ____

TUESDAY Number Sense and Operations

1. How many tens and ones?

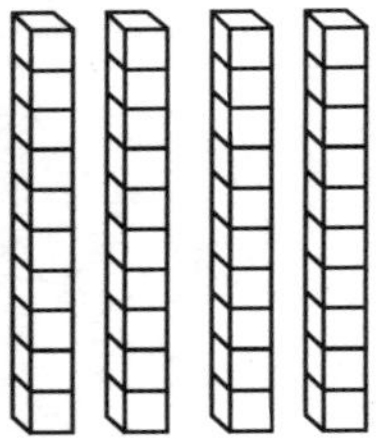

tens ______

ones ______

2. What is the value of the coins?

__________ ¢

3. What is the number between 19 and 21?

   19, ______, 21

4. Write the number.

   twelve ____________

WEDNESDAY Geometry

How many?  ______ 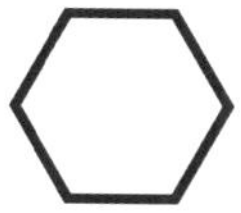 ______  ______

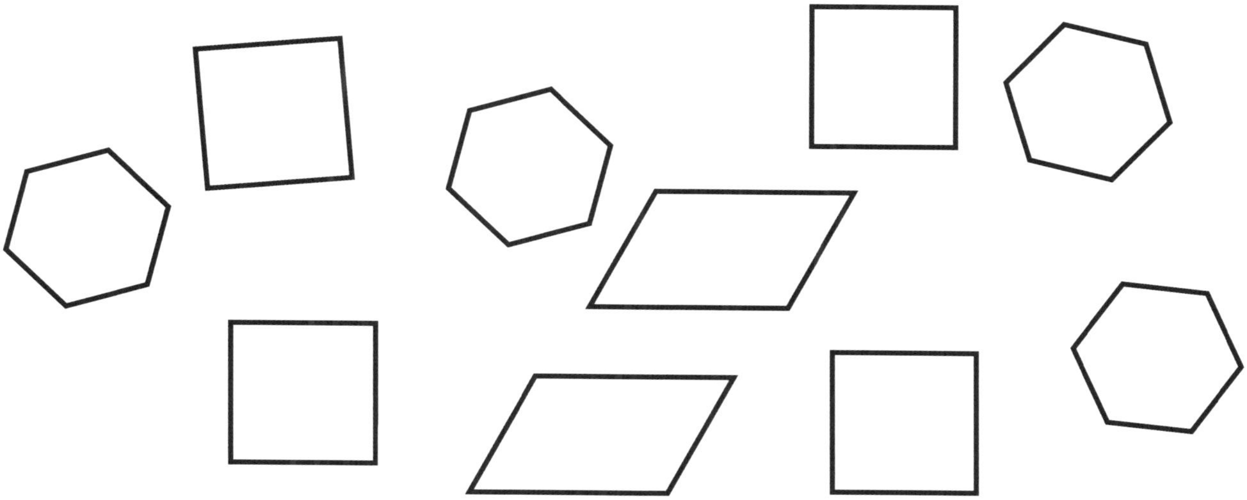

THURSDAY Measurement

1. What time is it?

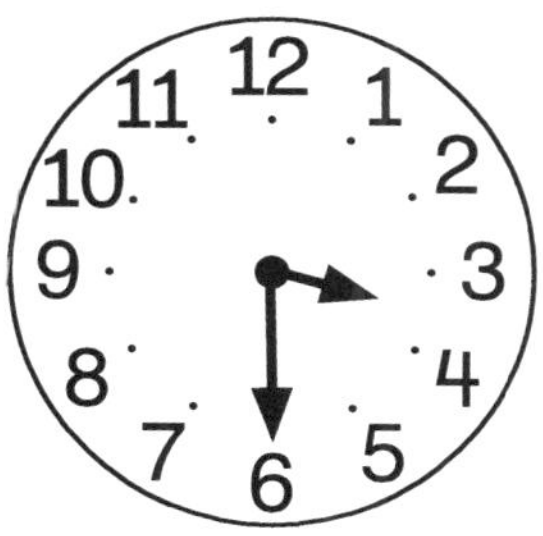

half past ______

2. Which fish is the shortest?

A.  B.  C. 

3. Measure the length of the line.

It is about ________  long.

FRIDAY Data Management

Favourite Activity

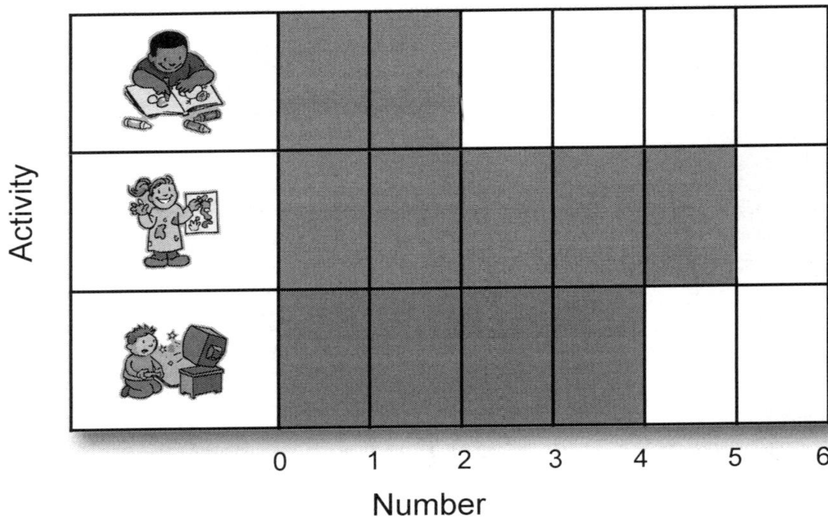

1. How many? 

2. Circle the most popular activity. 

3. Circle the least popular activity. 

BRAIN STRETCH

1. $\begin{array}{r} 3 \\ +\ 9 \\ \hline \end{array}$

2. $\begin{array}{r} 4 \\ +\ 6 \\ \hline \end{array}$

3. $\begin{array}{r} 10 \\ -\ \ 4 \\ \hline \end{array}$

4. $\begin{array}{r} 5 \\ -\ 1 \\ \hline \end{array}$

MONDAY Patterning and Algebra

1. Colour an AABB pattern. Circle the core of the pattern.

2. Cross out the blocks you want to take away.
   Colour the blocks left. Complete the subtraction sentence.

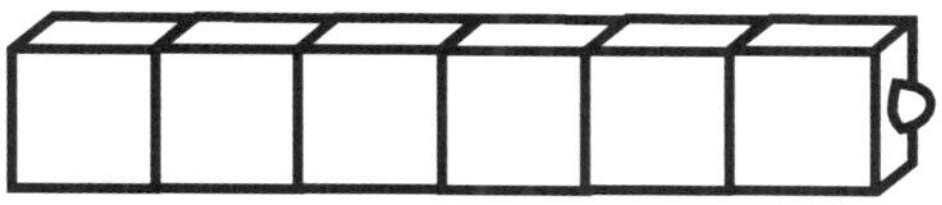

6 − ________ = ________

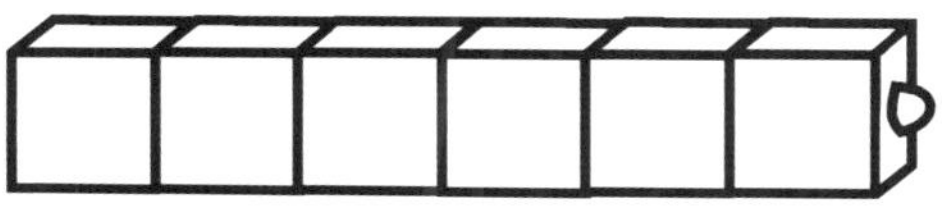

6 − ________ = ________

TUESDAY Number Sense and Operations

1. How many tens and ones?

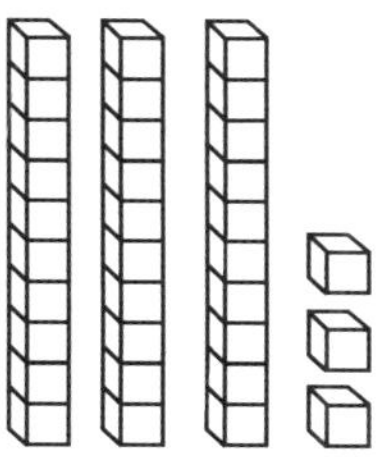

tens ______

ones ______

2. Circle the second monster.

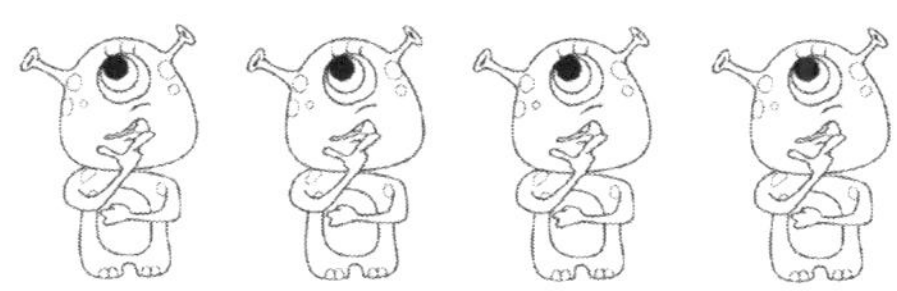

3. What is the number after 16?

15, 16, ______

4. Compare the numbers.

18 is greater than / less than / equal to 22

WEDNESDAY Geometry

Colour the trapezoid red.
Colour the square yellow.

Colour the triangle green.
Colour the rectangle blue.

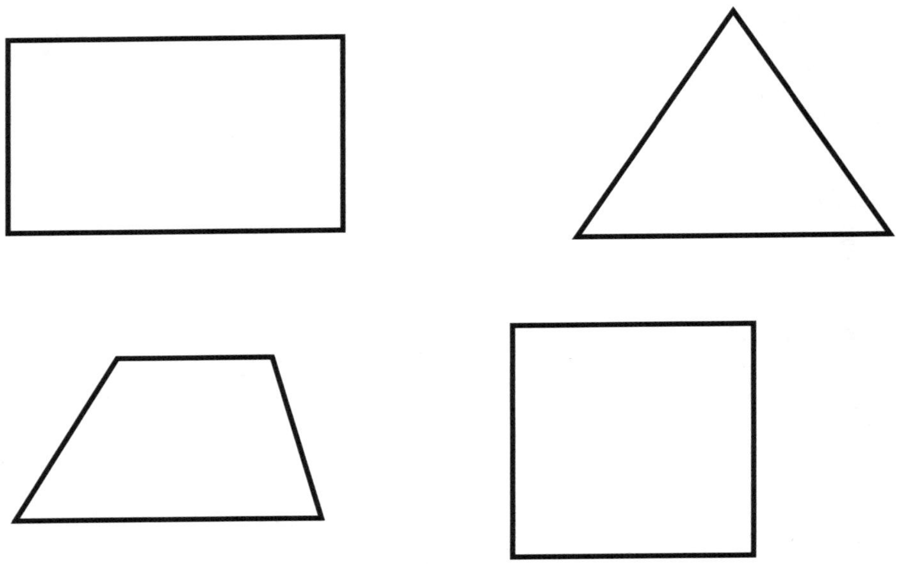

THURSDAY Measurement

1. What time is it?

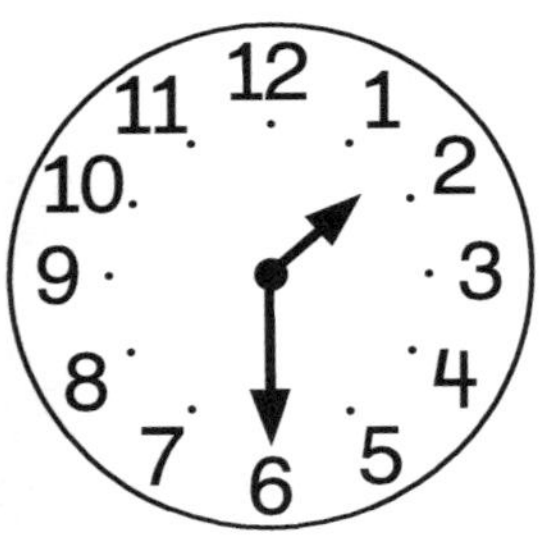

half past ______

2. Which container holds less?

A. 

B. 

3. Measure the length of the line.

It is about __________  long.

FRIDAY Data Management

Favourite Musical Activity

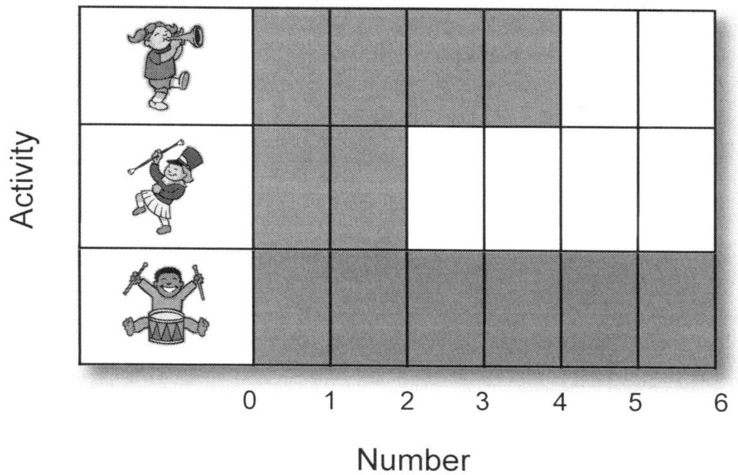

1. How many? 

2. Circle the most popular activity. 

3. Circle the least popular activity. 

BRAIN STRETCH

1. $6 + 5$

2. $5 + 5$

3. $12 - 4$

4. $6 - 3$

MONDAY Patterning and Algebra

1. Colour an AAB pattern. Circle the core of the pattern.

Add the doubles.

2. 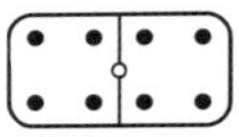 ________ + ________ = ________

3. 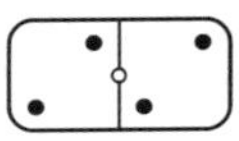 ________ + ________ = ________

4. 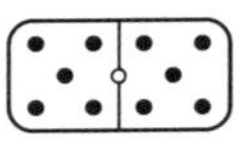 ________ + ________ = ________

TUESDAY Number Sense and Operations

1. How many tens and ones?

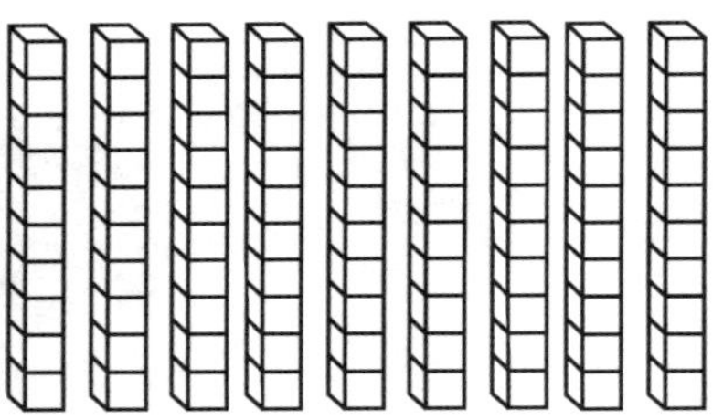

tens ______

ones ______

2. Write the numeral.

seventeen ______

3. Circle the fourth monster.

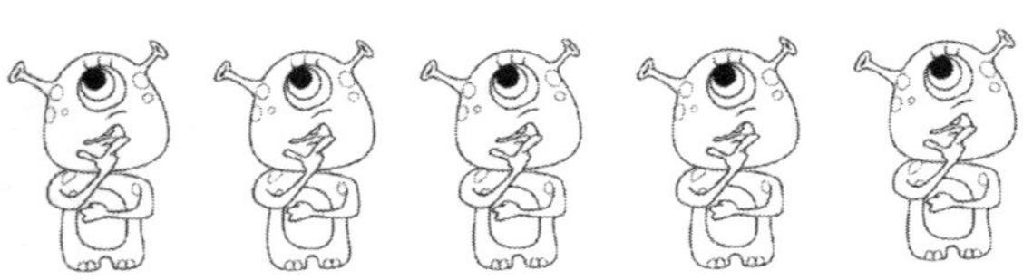

4. Colour one half of the shape.

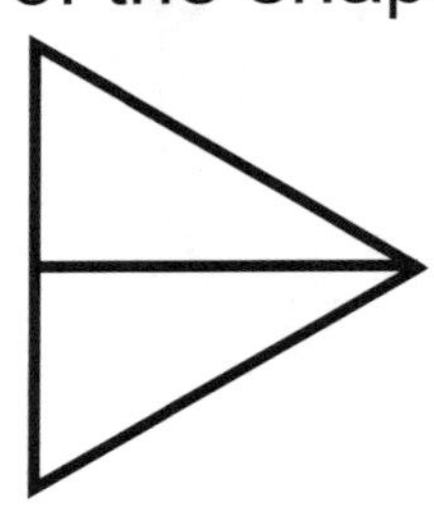

WEDNESDAY Geometry

1. Colour the shapes that are the same size and shape.

2. What is the name of this 3D shape?

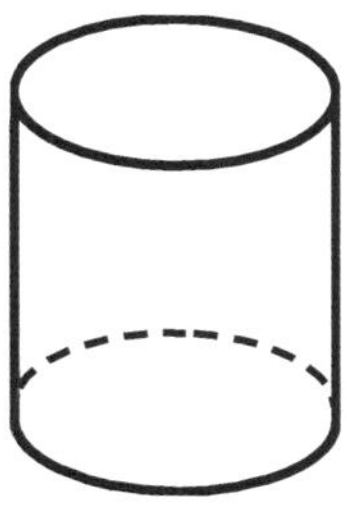

A. cylinder

B. pyramid

3. What shape is inside the circle?

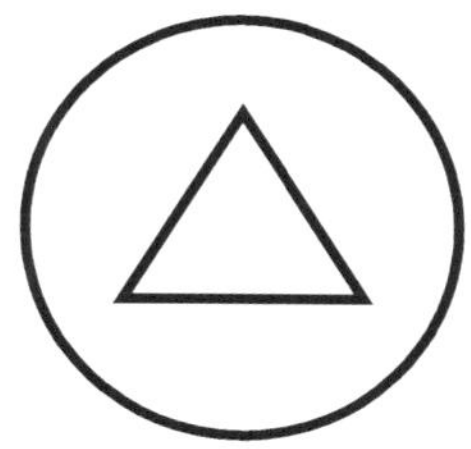

A. triangle

B. square

THURSDAY Measurement

1. What time is it?

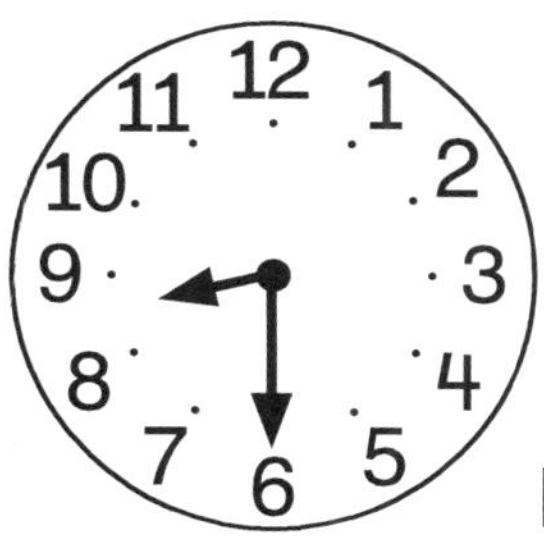

half past ______

2. Circle the best measurement tool for the temperature.

A. 

B. 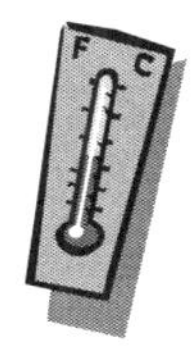

3. Measure the length of the line.

It is about __________  long.

FRIDAY Data Management

1. Count the pictures to complete the bar graph.

Favourite Fish Graph

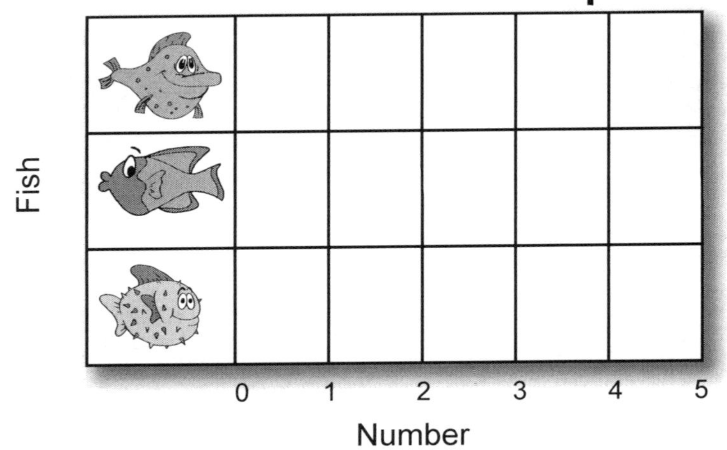

2. Circle the most popular fish. 

3. How many more  than 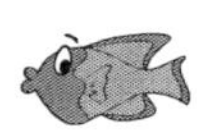 ? Write a number sentence.

__________ – __________ = __________

BRAIN STRETCH

Count on by 2s.

2 _____ _____ _____ _____ _____

MONDAY Patterning and Algebra

1. Tanya counts 3 red bugs, 4 blue bugs, and 2 green bugs. How many bugs are there in all?

3 + 4 + 2 = ________

2. There are 4 white chicks, 6 yellow chicks, and 1 orange chick. How many are there in all?

4 + 6 + 1 = ________

TUESDAY Number Sense and Operations

1. How many tens and ones?

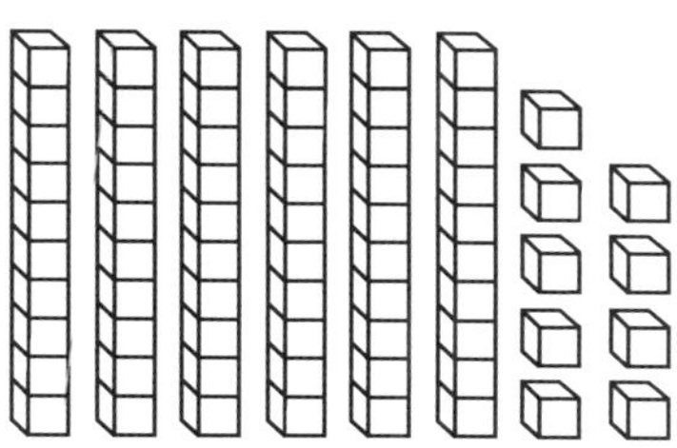

tens ______

ones ______

2. What is the value of the coins?

________ ¢

3. Order the numbers from smallest to largest.

11, 15, 13

______, ______, ______

4. Circle the fifth monster.

WEDNESDAY Geometry

1. Colour the shapes that are the same size and shape.

2. What is the name of this 3D shape?

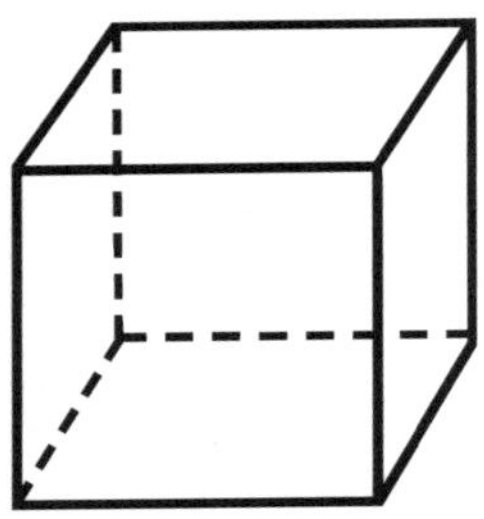

A. cylinder

B. cube

3. What shape is inside the triangle?

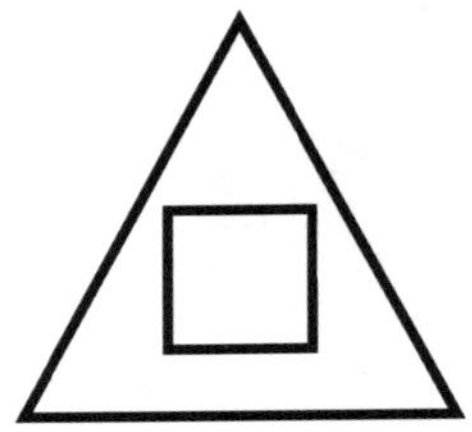

A. circle

B. square

THURSDAY Measurement

1. What time is it?

half past ______

2. Circle the best measurement tool to find out the date.

A. 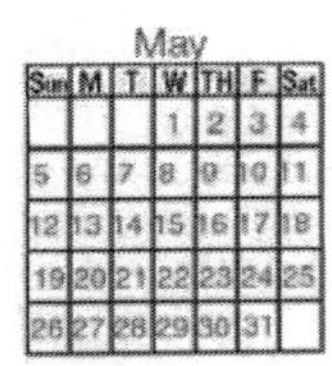

B. 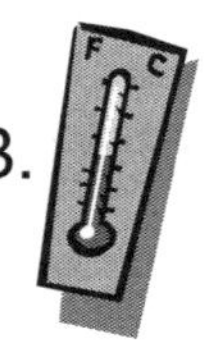

3. Measure the length of the line.

It is about __________  long.

FRIDAY Data Management

1. Count the pictures to complete the bar graph.

Favourite Pet Graph

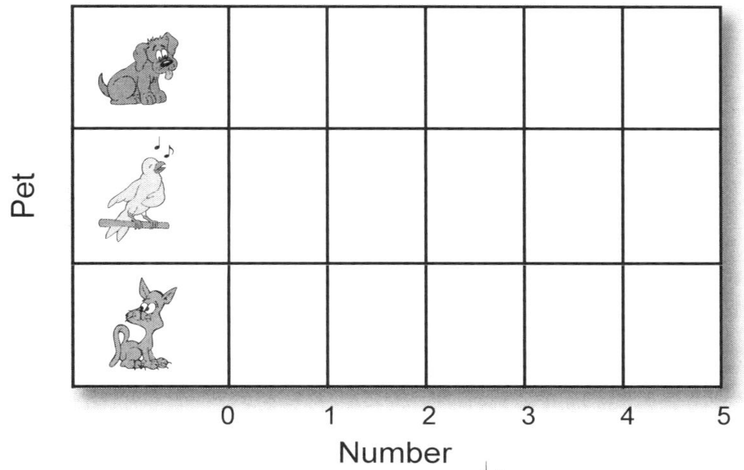

2. How many?  ________  ________  ________

3. How many more  than  ? Write a number sentence.

__________ – __________ = __________

BRAIN STRETCH

Count on by 5s.

5 _____ _____ _____ _____ _____

MONDAY Patterning and Algebra

1. There are 5 red birds, 4 blue birds, and 4 green birds. How many are there in all?

5 + 4 + 4 = ________

2. Ava counted 5 red flowers, 7 yellow flowers, and 2 pink flowers. How many are there in all?

5 + 7 + 2 = ________

TUESDAY Number Sense and Operations

1. How many tens and ones?

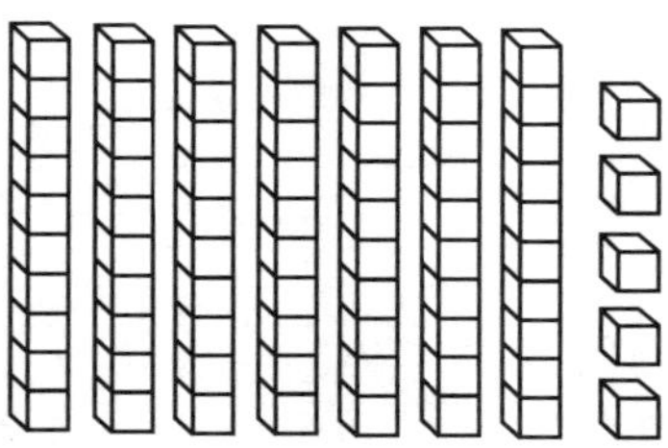

tens _____

ones _____

2. What is the value of the coins?

________ ¢

3. Order the numbers from largest to smallest.

13, 18, 11

_____, _____, _____

4. Colour one half of the shape.

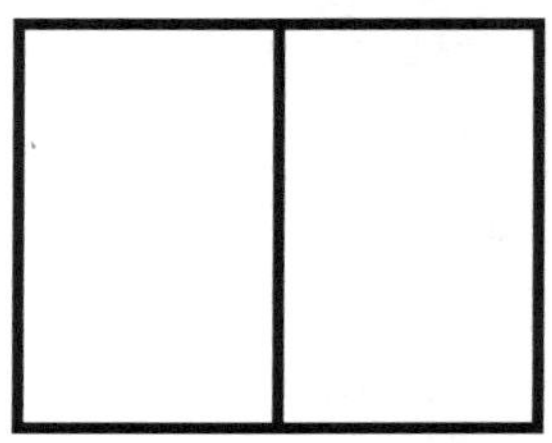

WEDNESDAY Geometry

1. Colour the hexagons red. Colour the circles blue.
   Colour the triangles yellow. Colour the rectangle green.

2. What is the name of this 3D shape?

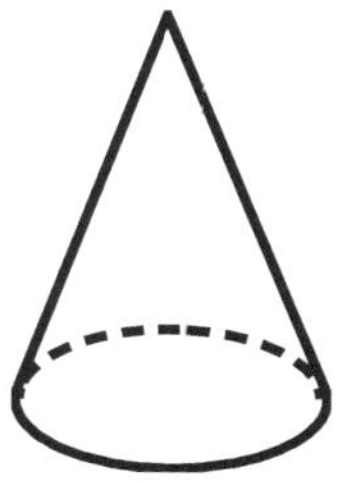

A. cone

B. pyramid

3. I am a closed figure. I have 3 sides. What shape am I?

 ______________

THURSDAY Measurement

1. What time is it?

half past ______

2. Circle the best measurement tool for the length of a book.

A. 

B. 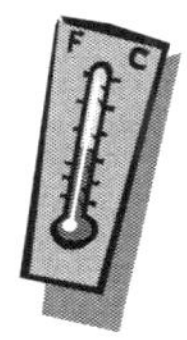

3. Measure the length of the line.

It is about __________  long.

FRIDAY Data Management

1.Count the pictures to complete the bar graph.

Favourite Treat Graph

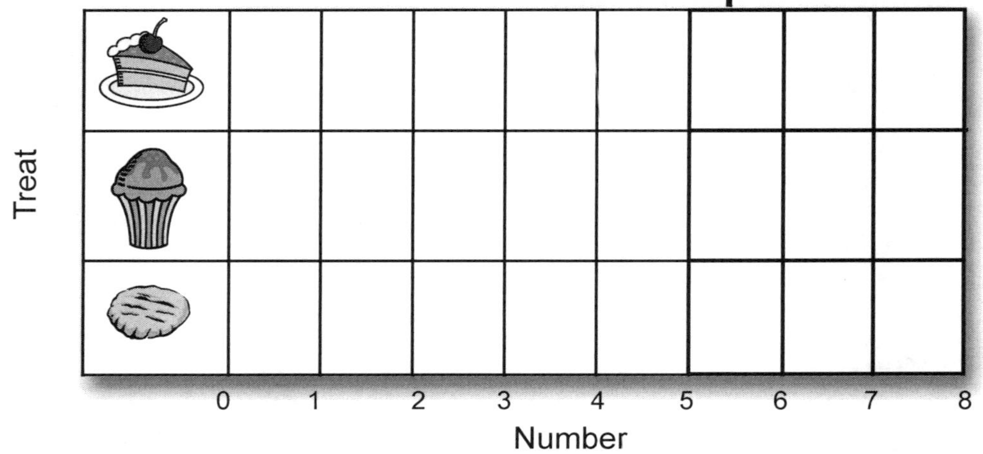

2. How many? 

3. How many fewer 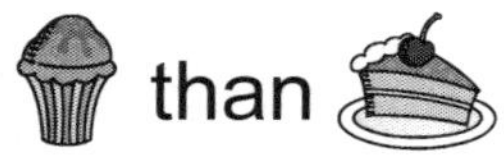 than ? Write a number sentence.

_________ – _________ = _________

BRAIN STRETCH

How many groups of 10? _______ How many left over? _______

Write the number. ______

MONDAY Patterning and Algebra

1. Create a pattern where shape and size changes.

____

2. $5 + 2 =$ ________

3. $9 - 1 =$ ________

4. $3 + 7 =$ ________

5. $7 - 6 =$ ________

TUESDAY Number Sense and Operations

1. How many tens and ones?

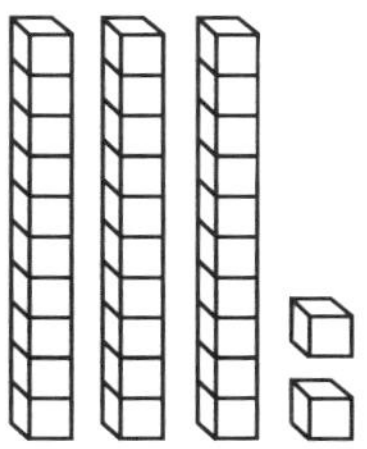

tens ______

ones ______

2. Write the number.

nineteen ______

3. Order the numbers from largest to smallest.

20, 14, 12

______, ______, ______

4. Colour one fourth of the shape.

WEDNESDAY Geometry

1. Colour the trapezoids red.
   Colour the pentagons yellow.
   Colour the square green.
   Colour the rectangles red.

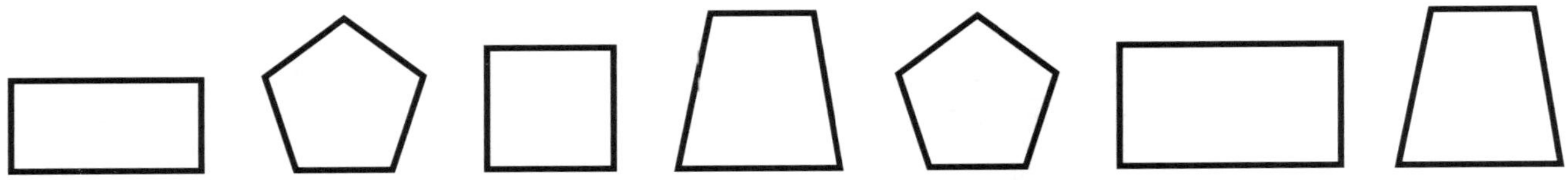

2. What is the name of this 3D shape?

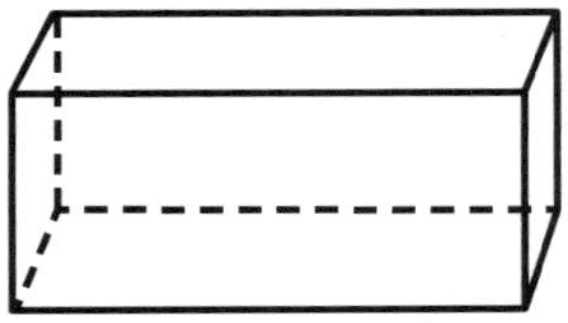

A. rectangular prism

B. cone

3. I am a closed figure. I have 4 sides all the same length. What shape am I?

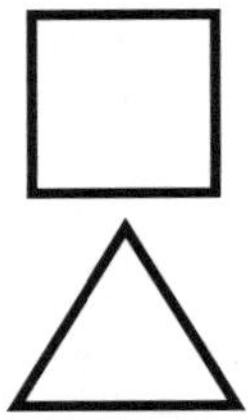

____

THURSDAY Measurement

1. What time is it?

half past ______

2. Circle the best measurement tool for the mass of some bananas.

A. 

B. 

3. Measure the length of the line.

It is about __________ 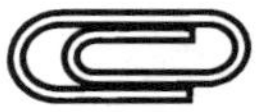 long.

FRIDAY Data Management

1. Count the pictures to complete the bar graph.

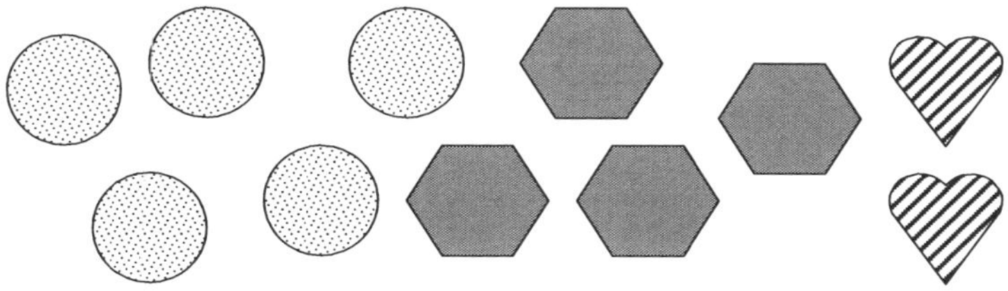

Favourite Shape Graph

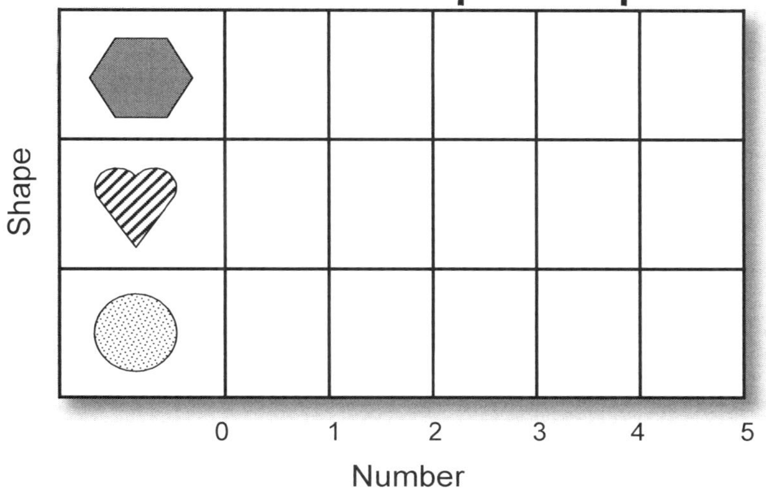

2. How many shapes altogether? 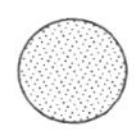 +  + 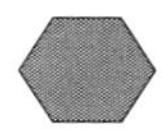 = ______

BRAIN STRETCH

Show two ways to add 4 + 3 + 6. Use the number lines.

4 + 3 + 6 = ____

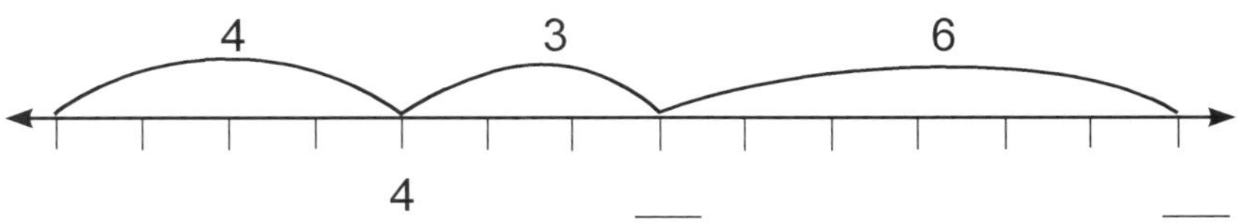

4 + 6 + 3 = ____

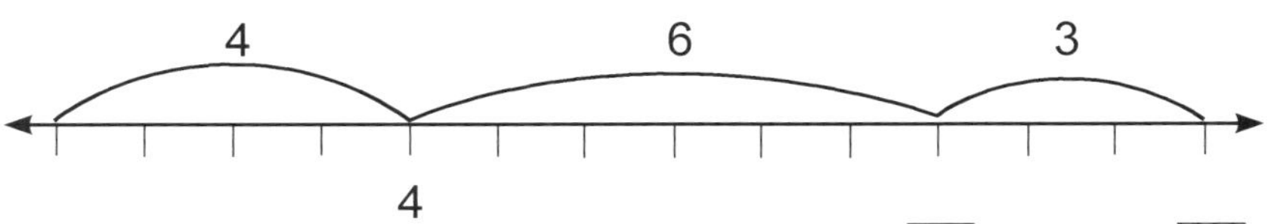

MONDAY Patterning and Algebra

1. Count out 7 counters. Add 7 more counters.
   Add 1 more counter. Count all the counters.

   7 + 7 + 1 = ________

2. Use counters to add. 8 + 8 + 1 = ________

3. What goes with 2 to make 11?

   2
   + ☐
   ———
   11

4. What goes with 6 to make 12?

   ☐
   + 6
   ———
   12

TUESDAY Number Sense and Operations

1. How many tens and ones?

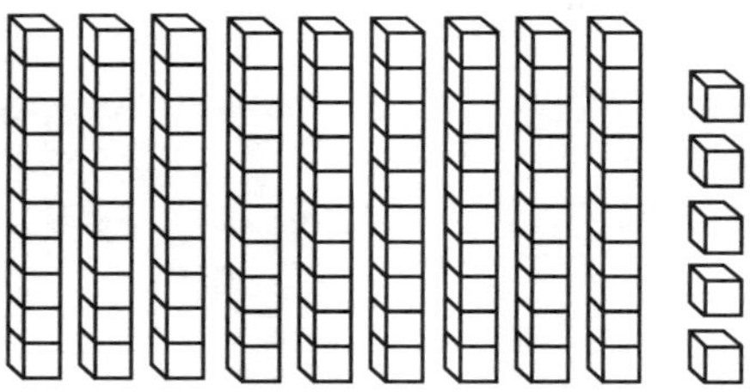

   tens _____

   ones ______

2. Write the number.

   eleven ______

3. Order the numbers from largest to smallest.

   30, 11, 26

   ______, ______, ______

4. Add 35 + 10.
   Show your work on the number line.

   35 + 10 = ____

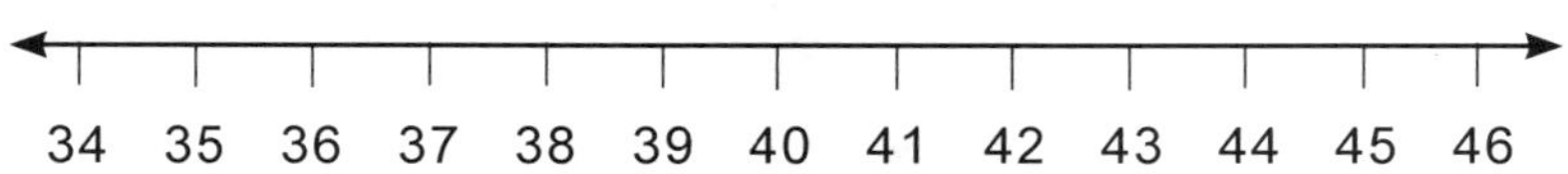

WEDNESDAY Geometry

1. Colour the pentagons red.
   Colour the triangles blue.
   Colour the trapezoids orange.
   Colour the circles green.

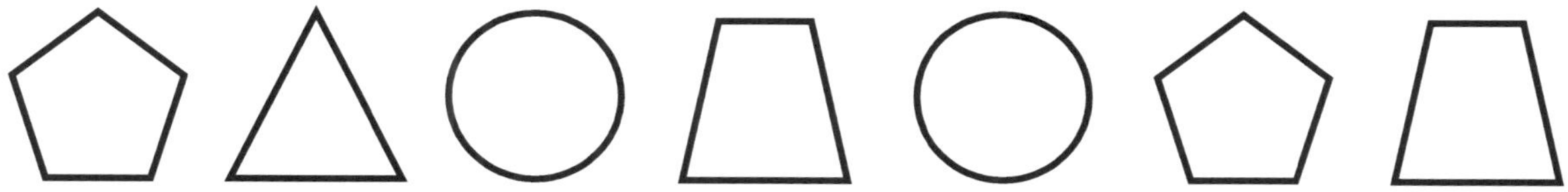

2. What is the name of this 3D shape?

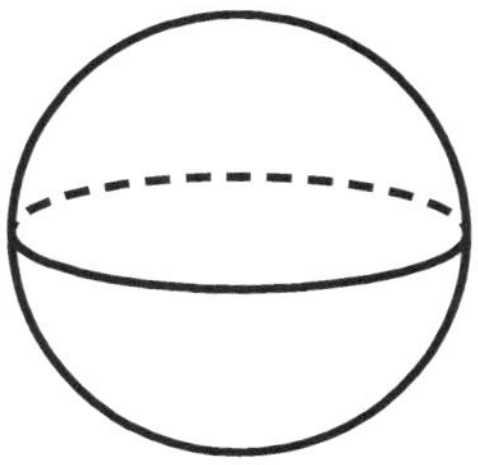

A. sphere

B. cone

3. I have 4 sides. My opposite sides are the same length. What shape am I?

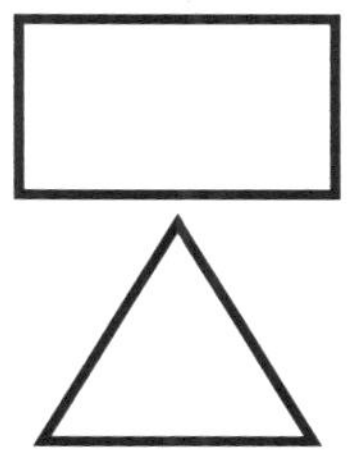

____

THURSDAY Measurement

1. What time is it?

half past ______

2. Circle the best measurement tool for a liquid.

A. 

B. 

3. Measure the length of the line.

It is about __________  long.

FRIDAY Data Management

1. Count the pictures to complete the bar graph.

Favourite Treat

2. How many?  ________  ________  ________

BRAIN STRETCH

How many groups of 10? ________ How many left over? ________

Write the number. _______

MONDAY Patterning and Algebra

1. Create a pattern. Circle the core of the pattern.

2. Ed has 3 red cards, 7 green cards, and 8 blue cards.
   How many cards does he have in all?

   ___ + ___ + ___ = ☐

   Ed adds 3 + 7 = ___ . Then he adds 8.

   So, there are ___ cards.

TUESDAY Number Sense and Operations

1. How many tens and ones?

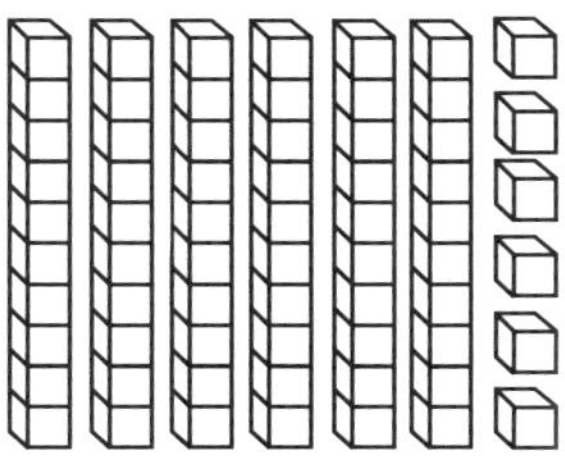

tens ______

ones ______

2. What is the value of the coins?

+

= ________ ¢

3. Write the numeral.

three ________

4. The cookie is cut in half. Each part is ______ half.

   There are ______ halves in all.

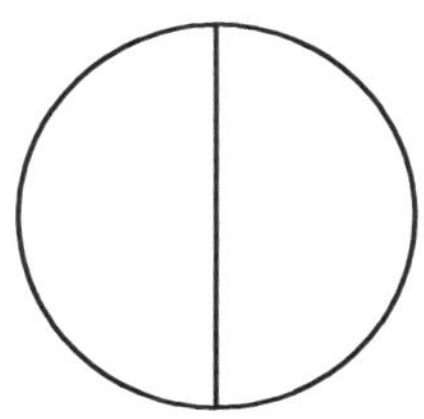

WEDNESDAY Geometry

1. Colour the shapes with more than 3 sides red.

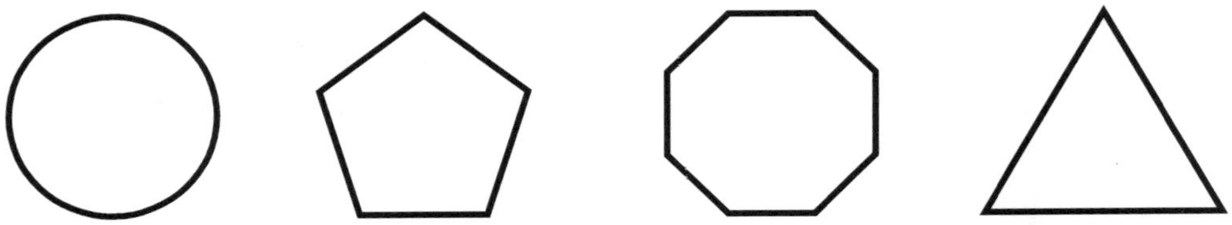

2. What is the name of this 3D shape?

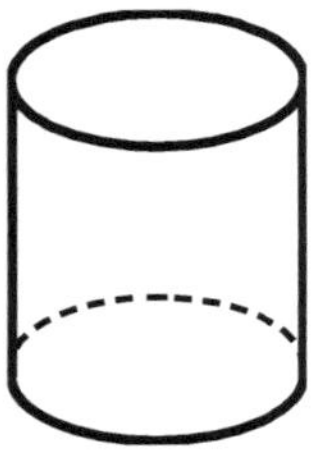

A. cone

B. cylinder

3. What shape is inside the circle?

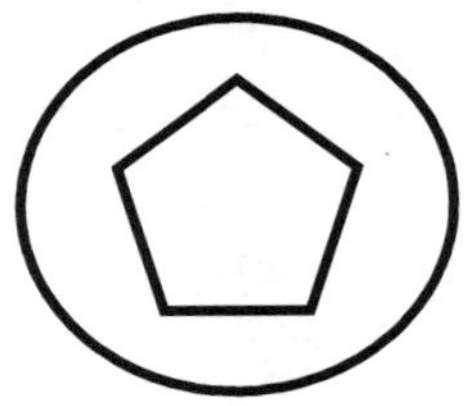

A. pentagon

B. rectangle

THURSDAY Measurement

1. What time is it?

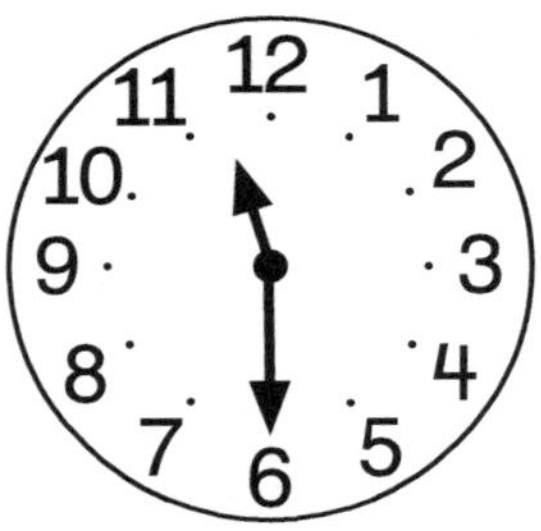

half past ______

2. How many months in a year?

____

3. Measure the length of the line.

It is about ___________  long.

FRIDAY Data Management

1. Complete the tally chart.

Favourite Pet

	Number	Tally
	5	
	3	
	7	

2. Circle the most popular pet. 

3. Circle the least popular pet. 

BRAIN STRETCH

Cam has 9 nut bars, 2 fruit bars, and 4 raisin bars.
How many bars does she have? Use the 10 box to add.

So, 9 + 4 + 2 = ____

Cam has __ bars.

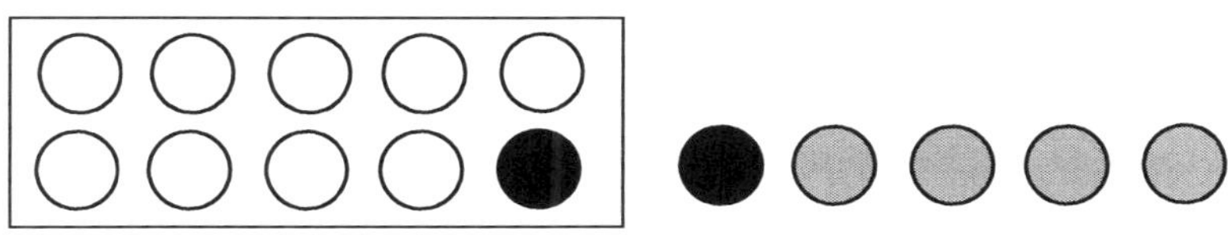

MONDAY Patterning and Algebra

1. Colour an ABB pattern. Circle the core of the pattern.

2. 2 + 7 = _______

3. 10 − 9 = _______

4. 11 + 1 = _______

5. 3 − 3 = _______

TUESDAY Number Sense and Operations

1. How many tens and ones?

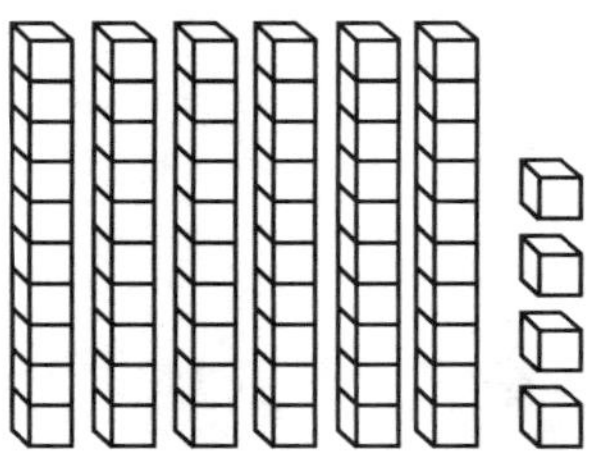

tens _____

ones ______

2. What is the value of the coins?

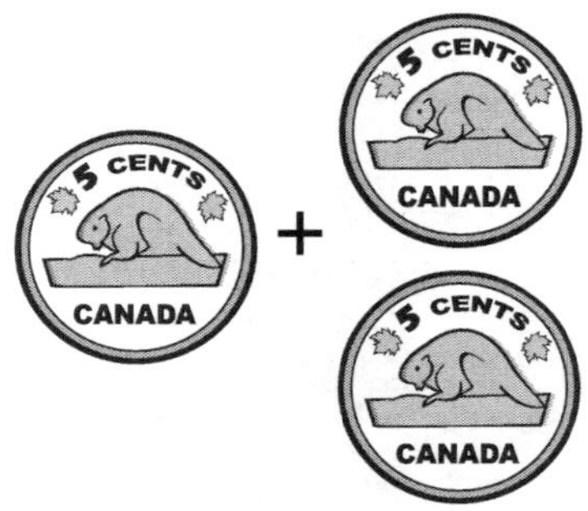

= ________ ¢

3. Write the numeral.

zero __________

4. Colour 1/3 of the shape.

WEDNESDAY Geometry

1. Colour the shapes with 4 vertices red.

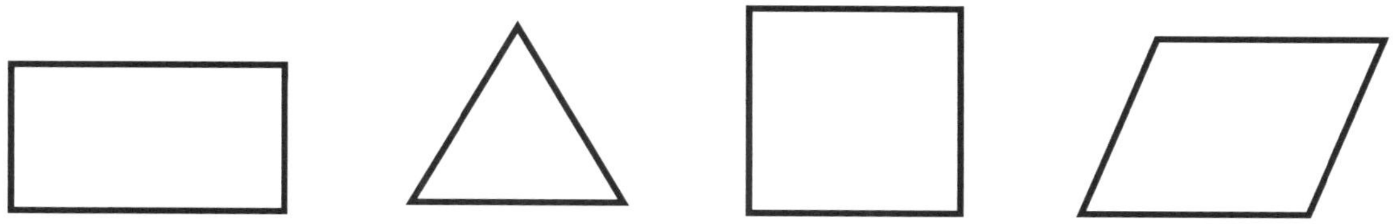

2. Can this 3D shape be stacked?

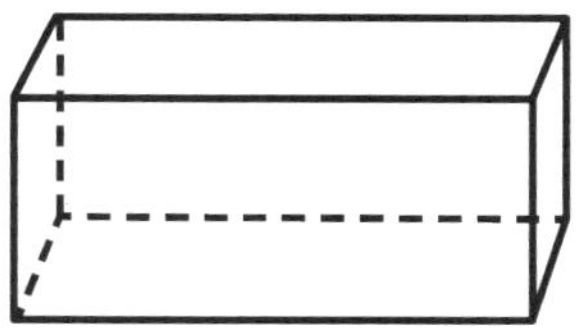

A. yes

B. no

3. What shape is inside the circle?

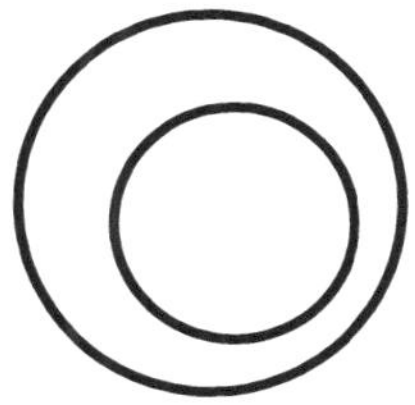

A. circle

B. rectangle

THURSDAY Measurement

1. What time is it?

half past ______

2. Which rectangle is longest?

A.

B.

C.

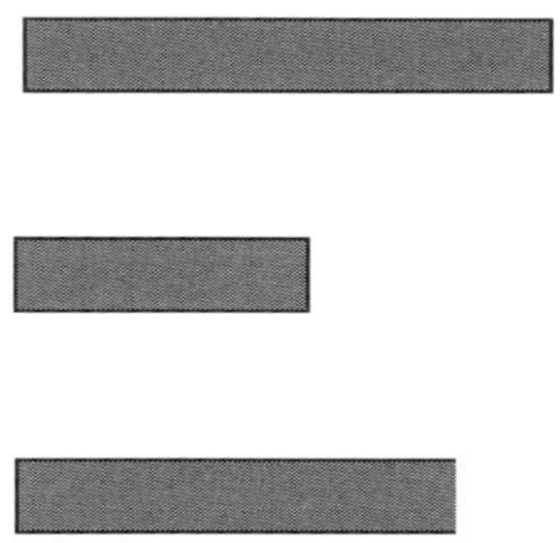

3. Measure the length of the line.

It is about __________  long.

FRIDAY Data Management

1. Complete the tally chart.

Favourite Fish

	Number	Tally
	6	
	4	
	5	

2. Circle the most popular fish.   

3. Circle the least popular fish.   

BRAIN STRETCH

There are 4  on the iceberg. Then 1  leaves.

How many  are there now?

______ ☐ ______ = ______

MONDAY Patterning and Algebra

1. Create a pattern where the size of the shapes change. Circle the core of the pattern.

Use the number line or counters to help answer the questions.

1 2 3 4 5 6 7 8 9 10 11 12 13 14 15 16

2. ☐ + 11 = 13

3. 10 − ☐ = 9

4. ☐ + 7 = 16

5. 9 − 3 = ☐

TUESDAY Number Sense and Operations

1. How many tens and ones?

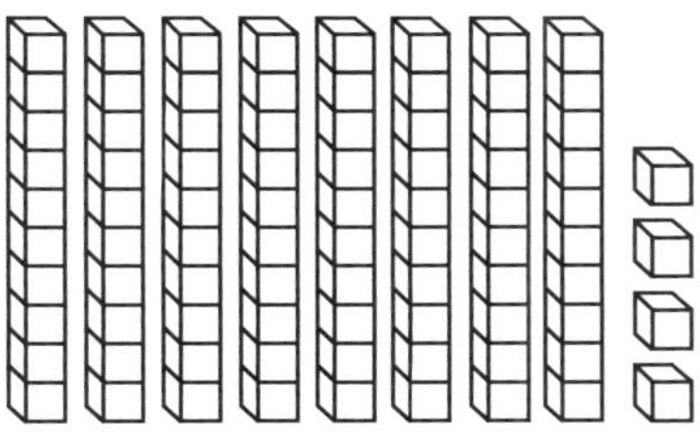

tens ______

ones ______

2. What number is 10 more than 50?

____

3. Write the numeral.

seven __________

4. The pizza is cut in fourths.
Each part is ___ fourth.
There are ___ fourths in all.

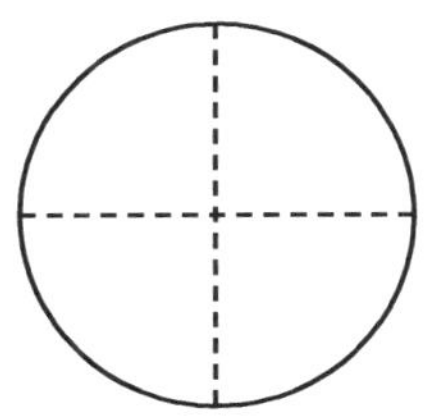

WEDNESDAY Geometry

1. Colour the shapes with less than 6 vertices red.

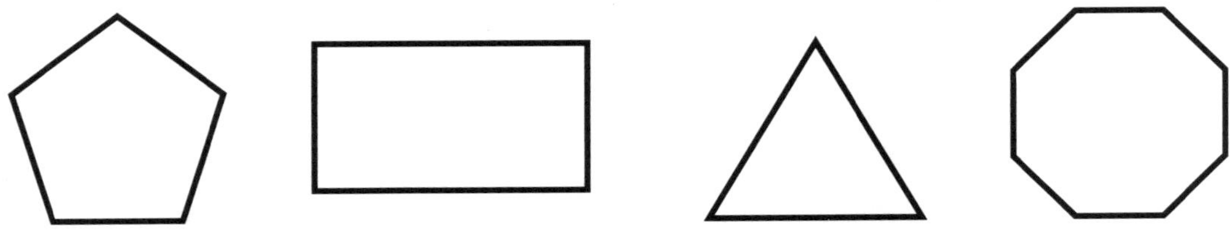

2. Can this 3D shape be stacked?

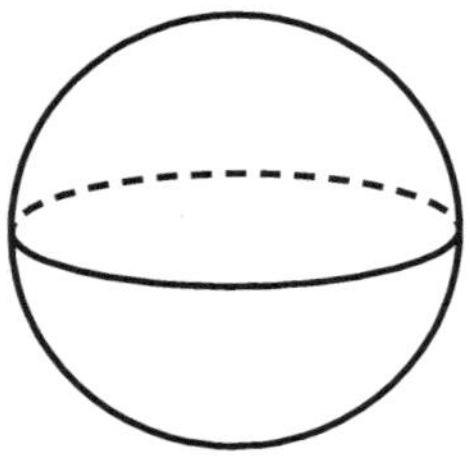

A. yes

B. no

3. What shape is beside the pentagon?

A. rectangle

B. triangle

THURSDAY Measurement

1. What time is it?

half past ______

2. A red bar is longer than a blue bar.

A blue bar is longer than a yellow bar.

Colour the bars in order from shortest to longest.

3. Measure the length of the line.

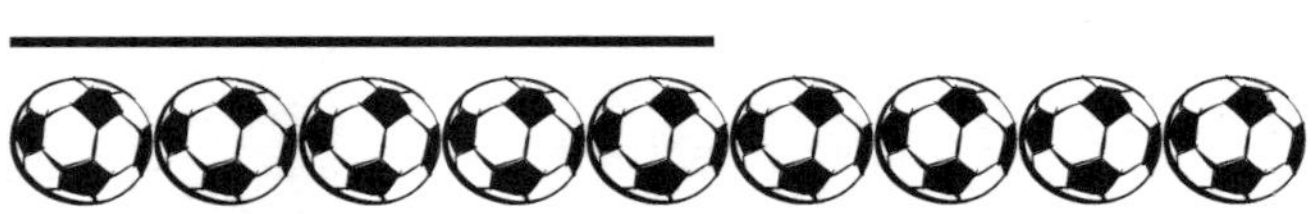

It is about __________  long.

FRIDAY Data Management

1. Complete the tally chart.

Favourite Playground Equipment

	Number	Tally
	2	
	8	
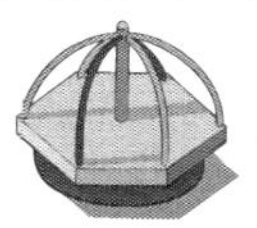	4	

2. Circle the most popular playground equipment.

  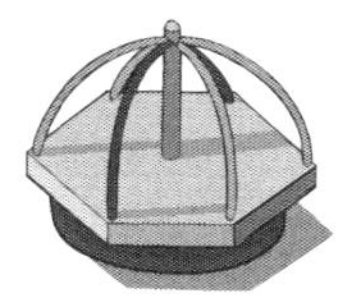

3. Circle the least popular playground equipment.

  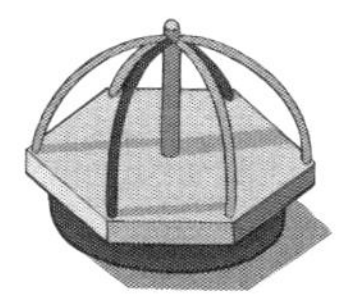

BRAIN STRETCH

Claire has 5 buttons. Jack has 8 buttons.
How many more buttons does Jack have than Claire?
Count out 5 counters. Then count on to 8.

5 + _______ = 8 Jack has _______ more buttons than Claire.

MONDAY Patterning and Algebra

1. Create a pattern where the size of the shapes change. Circle the core of the pattern.

2. 2 + 10 = ______

3. 3 − 1 = ______

4. 8 + 0 = ______

5. 8 − 4 = ______

TUESDAY Number Sense and Operations

1. How many tens and ones?

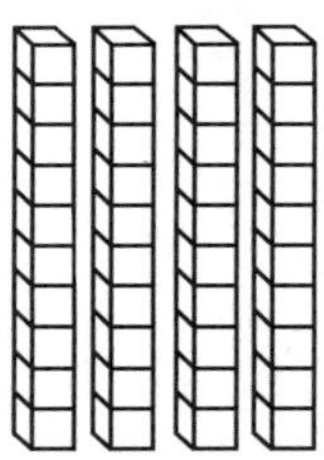

tens ______

ones ______

2. What number is 10 less than 17?

____

3. Write the numeral.

twenty __________

4. Draw a line to cut the rectangle into 2 equal parts.

Each part is ___ half.

There are ___ halves in all.

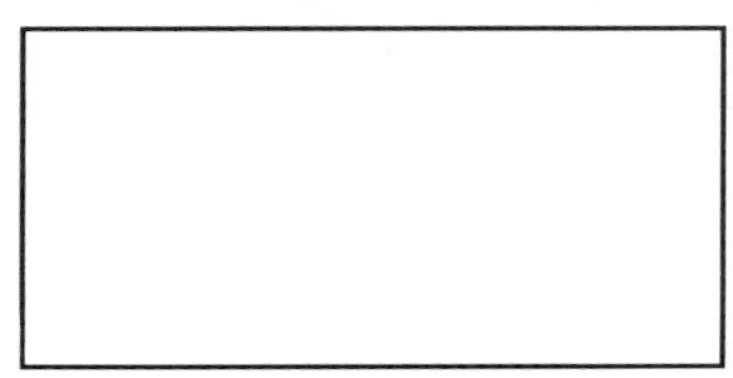

WEDNESDAY Geometry

1. Colour the shapes with more than 3 vertices blue.

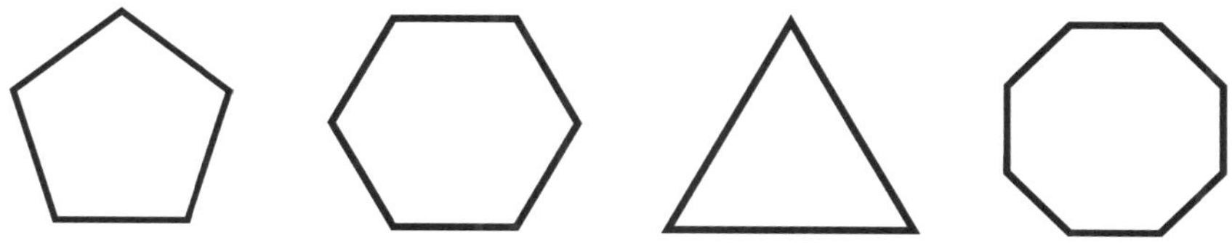

2. Can this shape roll?

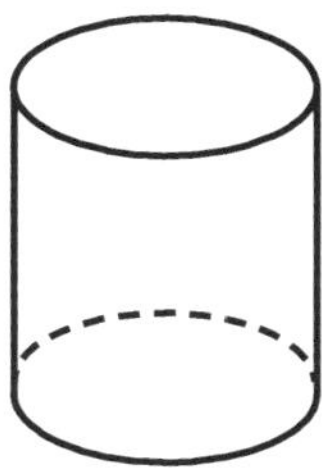

A. yes

B. no

3. What shape is inside the circle?

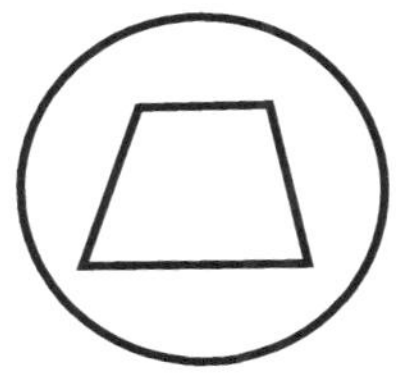

A. pentagon

B. trapezoid

THURSDAY Measurement

1. What time will it be 1 hour from now?

_____ : _____

2. Which rectangle is the longest?

A.

B.

C.

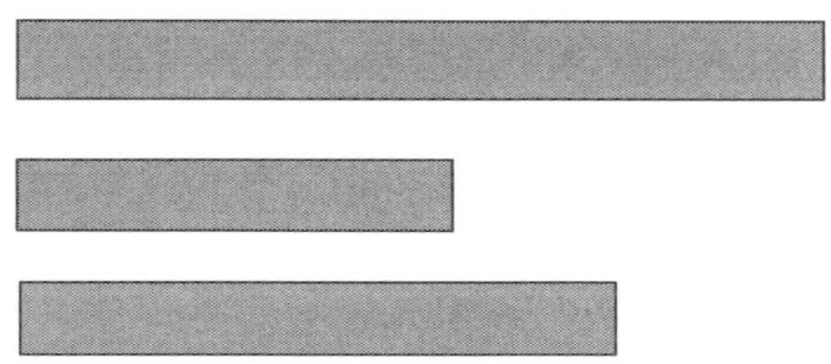

3. Which length is longer than the line?

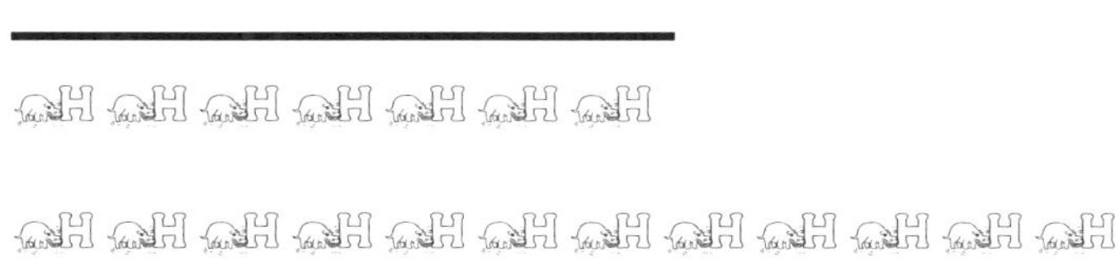

A.

B.

FRIDAY Data Management

1. Complete the tally chart.

Favourite Shape Graph

	Number	Tally
	2	
	7	
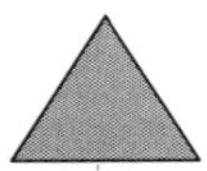	6	

2. Circle the most popular shape.

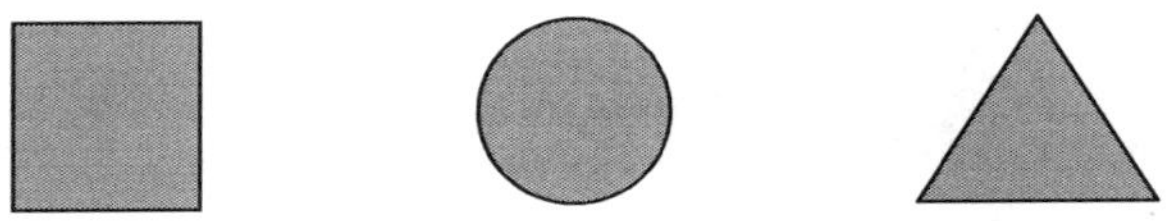

3. Circle the least popular shape.

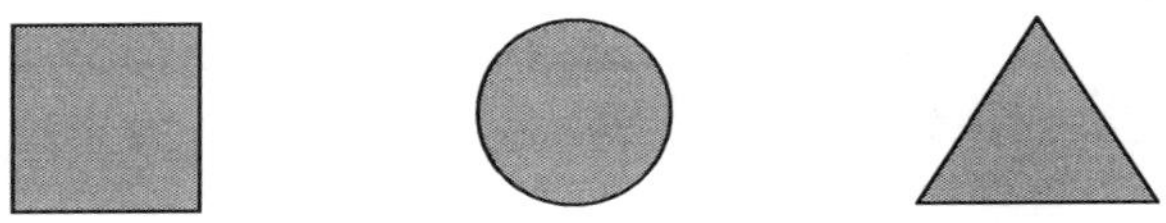

BRAIN STRETCH

Mandy has 7 pens. Mark has 10 pens.
How many fewer pens does Mandy have than Mark?

Mandy counts out 10 counters. Then she counts back to 7.

_____ – _____ = _____

Mandy has _____ fewer pens than Mark.

MONDAY Patterning and Algebra

1. Count back by 1s.

65, 64, 63, ______, ______, ______, ______

Use the number line or counters to help answer the questions.

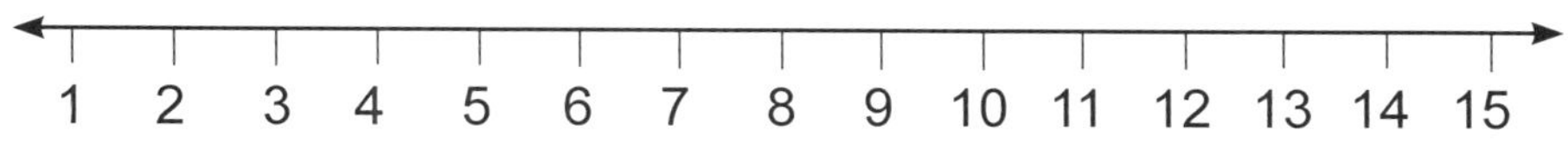

2. ______ + 1 = 14

3. 5 − ______ = 4

4. ______ + 5 = 15

5. 9 − 0 = ______

TUESDAY Number Sense and Operations

1. How many tens and ones?

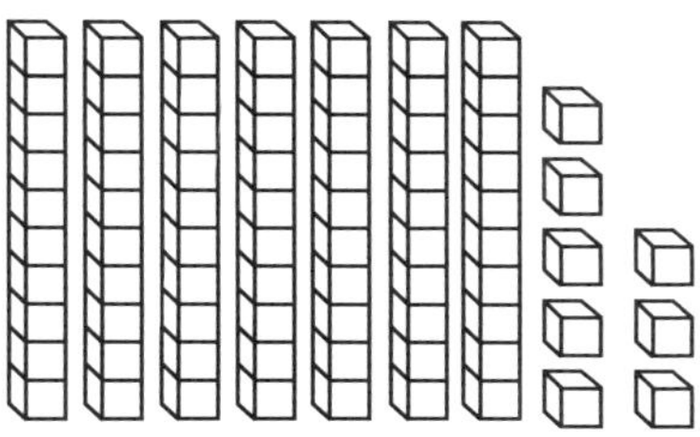

tens ______

ones ______

2. What number is 10 more than 2?

______

3. Write the numeral.

five ______

4. How many groups of 10? ______

How many left over? ______

Write the number. ______

WEDNESDAY Geometry

1. Colour the shapes with fewer than 5 sides green.

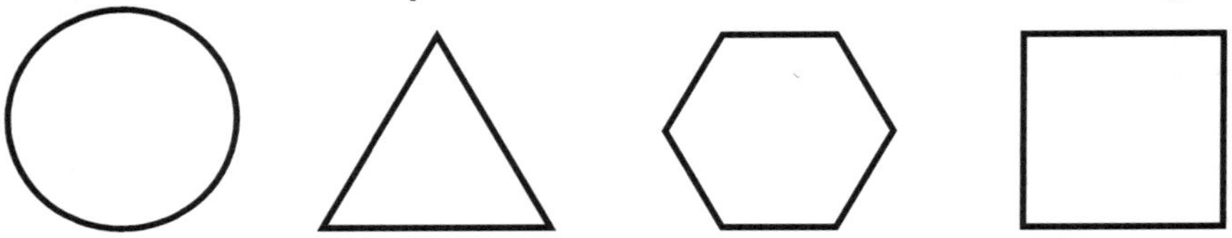

2. Can this 3D shape roll?

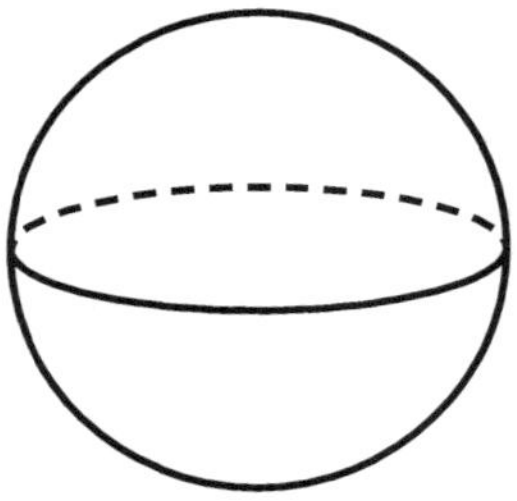

A. yes

B. no

3. What shape is inside the circle?

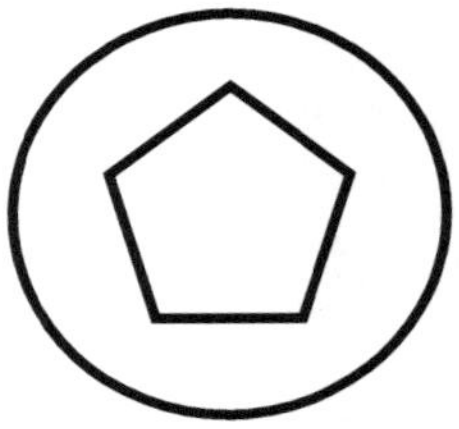

A. pentagon

B. rectangle

THURSDAY Measurement

1. What time will it be
   1 hour from now?

_____ : _____

2. Which rectangle is
   the shortest?

A.
B.
C.

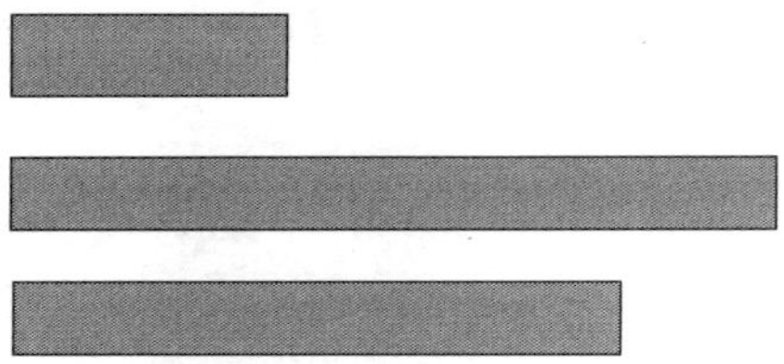

3. Which length is longer
   than the line?

A. 

B. 

FRIDAY Data Management

1. Complete the tally chart.

Favourite Activity

	Number	Tally
	8	
	6	
	2	

2. Circle the most popular activity. 

3. Circle the least popular activity. 

BRAIN STRETCH

There are 5 . Then 2 more  come.

How many  are there in all?

______ ☐ ______ = ______

MONDAY Patterning and Algebra

1. Count on by 2s.

22, 24, 26, ______, ______, ______, ______

Use the number line or counters to answer the questions.

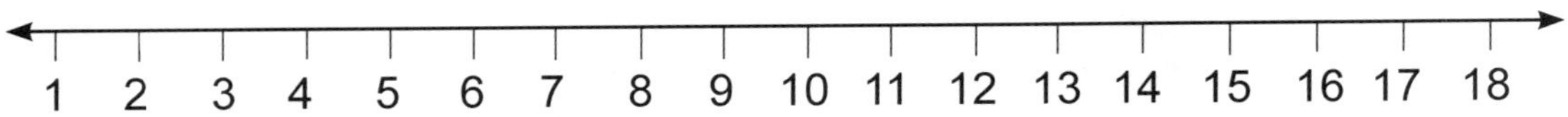

2. ☐ + 4 = 12

3. 16 − ☐ = 8

4. ☐ + 9 = 18

5. 8 − 7 = ☐

TUESDAY Number Sense and Operations

1. How many tens and ones?

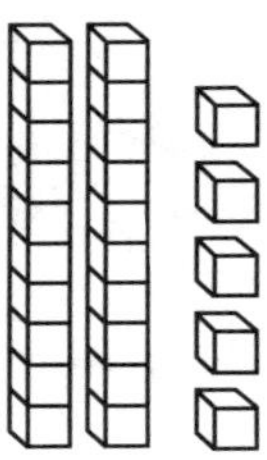

tens ______

ones ______

2. What number is 5 less than 12?

______

3. Write the numeral.

fourteen ______

4. Circle the correct sentence.

A. Each part is one half.

B. Each part is one quarter.

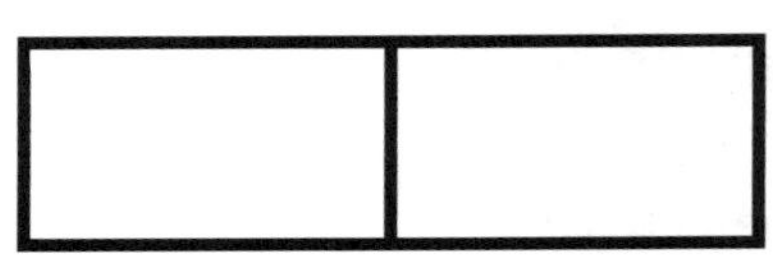

WEDNESDAY Geometry

1. What 3D shape can be made using these shapes?

A. 

B. 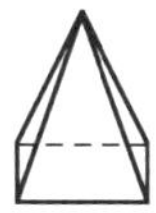

2. Draw a line of symmetry on the teddy bear.

3. Can this 3D shape roll?

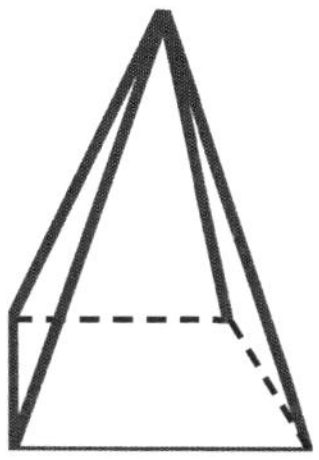

A. yes

B. no

4. What shape is under the circle?

A. rectangle

B. trapezoid

THURSDAY Measurement

1. What time will it be 1 hour from now?

half past ________

2. Which is lighter?

A. 

B. 

3. Which length is shorter than the line?

A. 

B. 

FRIDAY Data Management

1. Complete the tally chart.

Favourite Winter Activity

	Number	Tally
	5	
	3	
	9	

2. Circle the most popular winter activity.

3. Circle the least popular winter activity.

BRAIN STRETCH

Use the number line or counters to answer the questions.

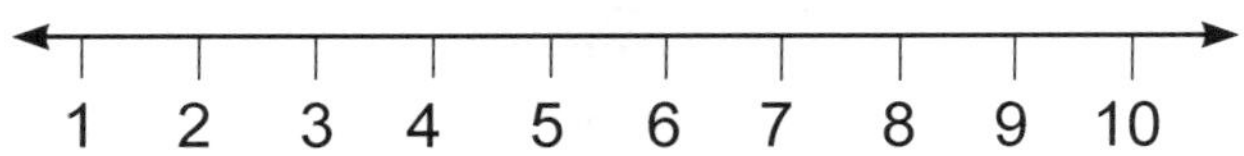

1. 5 + 2 = 3 + 4

True False

2. 6 – 3 = 2 + 1

True False

MONDAY Patterning and Algebra

1. Count on by 5s.

30, 35, 40, ________, ________, ________, ________

Use the number line or counters to answer the questions.

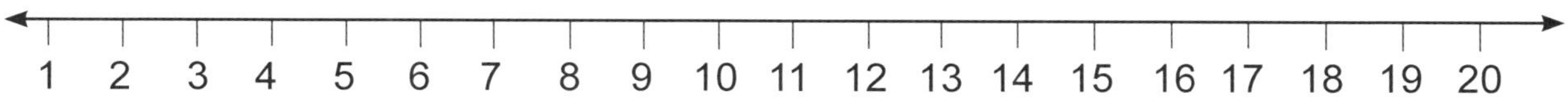

2. ________ + 10 = 11

3. 15 − ________ = 6

4. ________ + 3 = 10

5. 10 − 5 = ________

TUESDAY Number Sense and Operations

1. How many tens and ones?

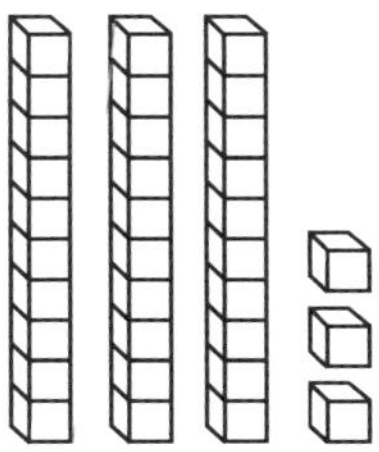

tens ______

ones ______

2. What number is 10 more than 40?

____

3. Write the numeral.

six ________

4. Draw lines to cut the circle into 4 equal parts.
Colour one fourth red.

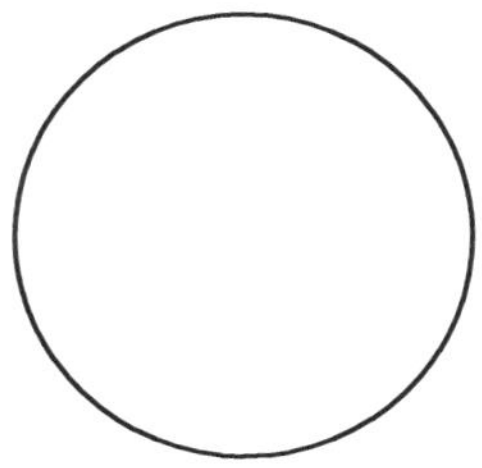

WEDNESDAY Geometry

1. Which shape shows a line of symmetry?

A.

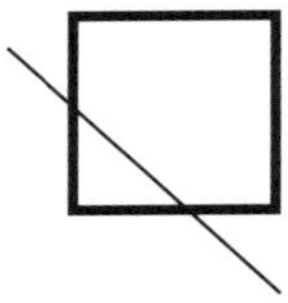

B.

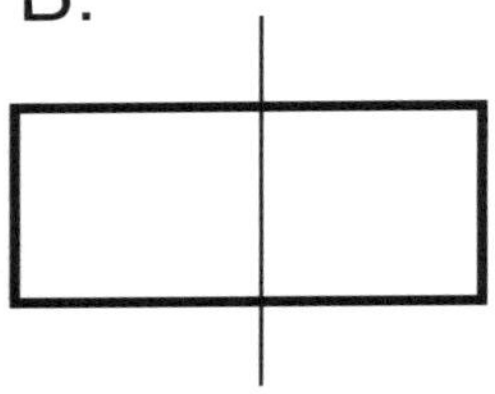

2. Can this 3D shape roll?

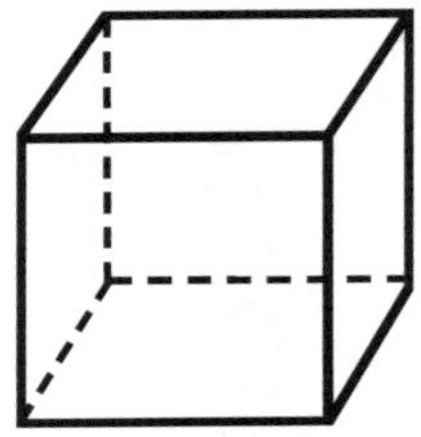

A. yes

B. no

3. What shape is above the pentagon?

A. circle

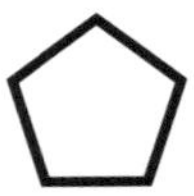

B. rectangle

THURSDAY Measurement

1. What time will it be 1 hour from now?

____ : ____

2. Which is heavier?

A.

B.

3. Both rows are the same length. Circle true or false.

A. True

B. False

FRIDAY Data Management

1. Complete the tally chart.

Favourite Fruit

	Number	Tally
	4	
	8	
	5	

2. Circle the most popular fruit.   

3. Circle the least popular fruit.   

BRAIN STRETCH

Use the number line or counters to answer the questions.

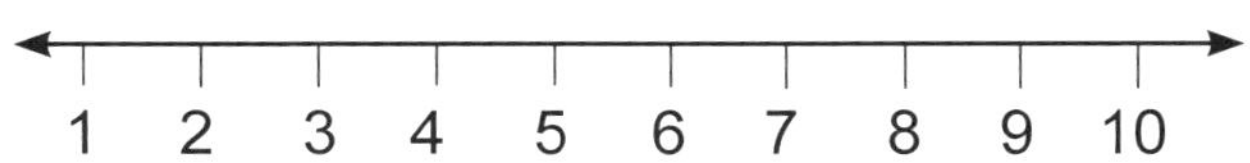

1. 7 + 1 = 4 + 4

True False

2. 8 – 4 = 2 + 1

True False

MONDAY Patterning and Algebra

1. Count on by 1s.

66, 67, 68, ________, ________, ________, ________

Use the number line or counters to answer the questions.

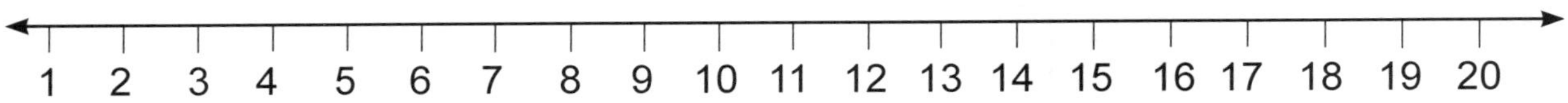

2. __________ + 7 = 14

3. 6 − __________ = 1

4. __________ + 10 = 20

5. 15 − 10 = __________

TUESDAY Number Sense and Operations

1. How many tens and ones?

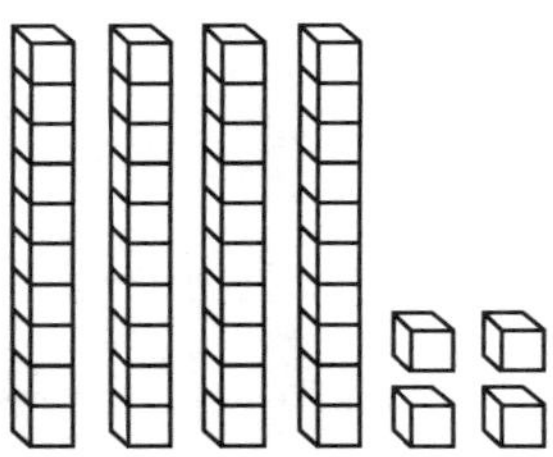

tens ______

ones _______

2. What number is 10 more than 62?

____

3. Write the numeral.

nine _________

4. What part of the circle is shaded?

____

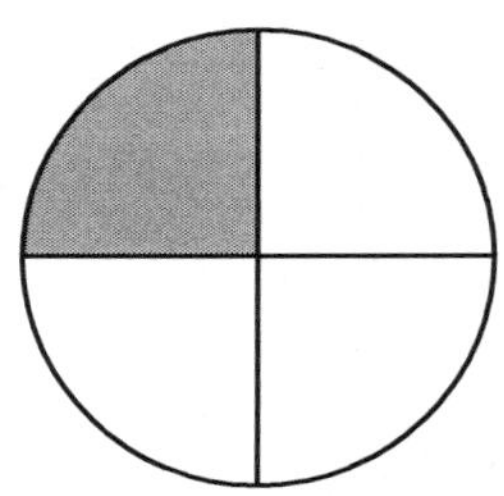

WEDNESDAY Geometry

1. What new shape can you make with 4 triangles?

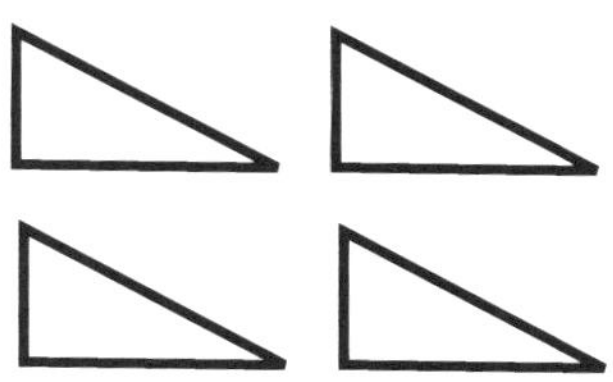

A. 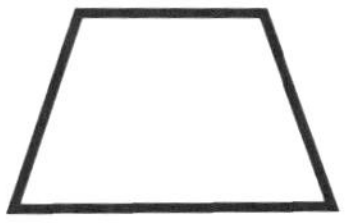

B. 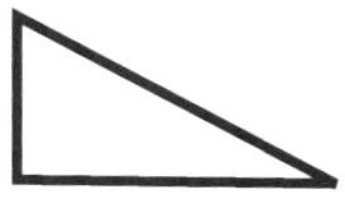

2. Which shape shows a line of symmetry?

A. 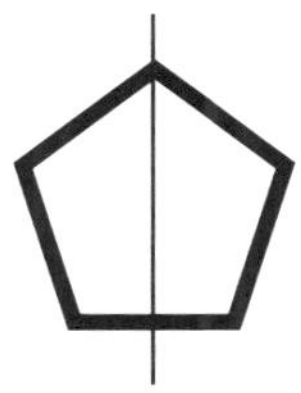

B. 

3. Tell how you know this shape is a trapezoid.

____

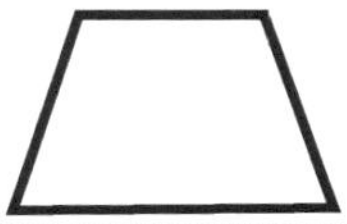

4. How many vertices does this shape have?

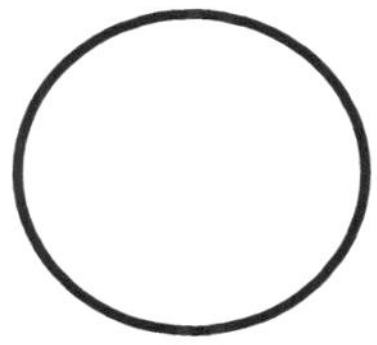

____

THURSDAY Measurement

1. What time will it be 1 hour from now?

____ : ____

2. Which holds less?

A. 

B. 

2. Is one row longer? Explain.

A. ____

B. 

FRIDAY Data Management

1. Complete the tally chart.

Favourite Toy

	Number	Tally
	2	
	6	
	8	

2. Circle the most popular toy.   

3. Circle the least popular toy.   

BRAIN STRETCH

Make two correct number sentences using subtraction.

3 7 4

_______ – _______ = _______

_______ – _______ = _______

MONDAY Patterning and Algebra

1. Count on by 10s.

30, 40, 50, 60, _______, _______, _______, _______

Use the number line or counters to answer the questions.

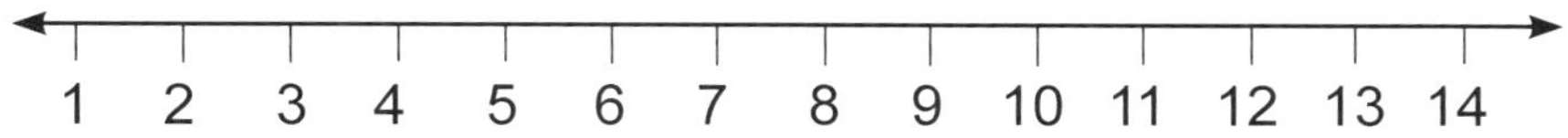

2. _________ + 5 = 10

3. 11 − _________ = 7

4. _________ + 9 = 14

5. 11 − 3 = _________

TUESDAY Number Sense and Operations

1. How many tens and ones?

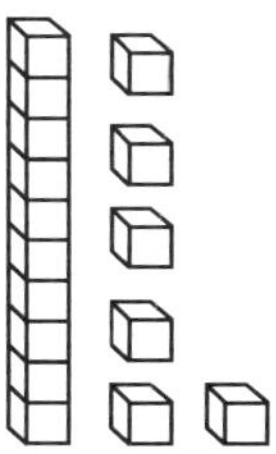

tens ______

ones _______

2. What is the value of the coins?

= _________ ¢

3. Order the numbers from largest to smallest.

23, 3, 33

______, ______, ______

4. Circle the piece that is smaller.

A. one half

B. one fourth

WEDNESDAY Geometry

Colour the 3D shapes that have only flat surfaces red.
Colour the 3D shapes that have only curved surfaces blue.
Colour the 3D shapes that have both types of surfaces green.

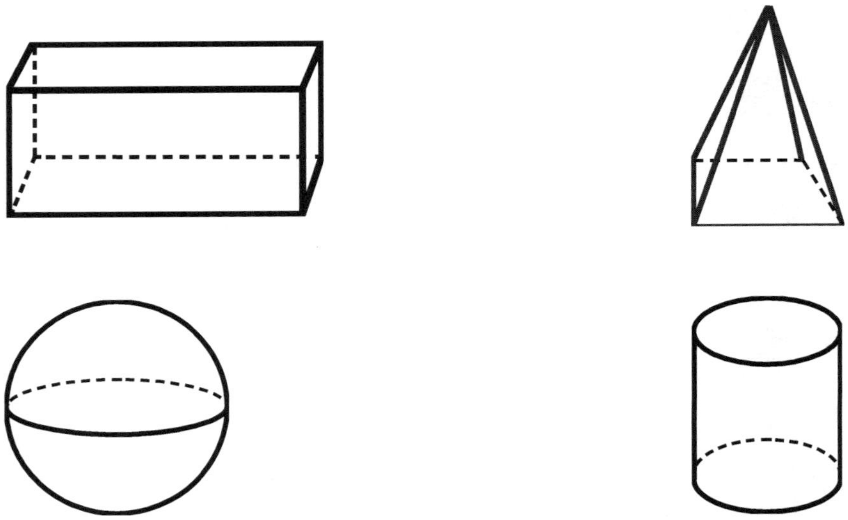

THURSDAY Measurement

1. What time will it be 1 hour from now?

____ : ____

2. Which container holds more?

A. 

B. 

3. Measure the length of the line.

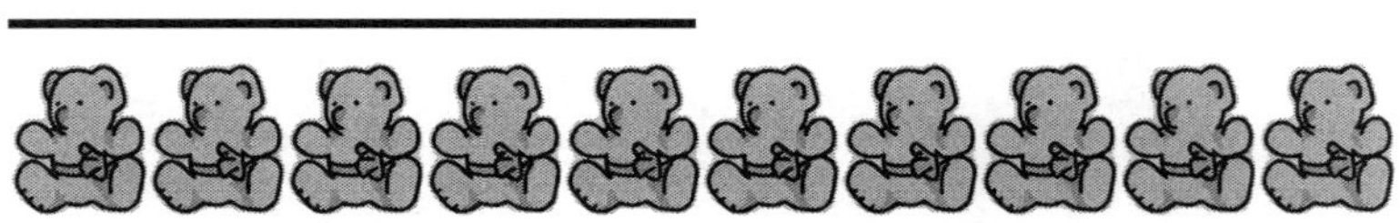

It is about __________  long.

FRIDAY Data Management

1. Complete the tally chart.

Favourite Vegetable

	Number	Tally
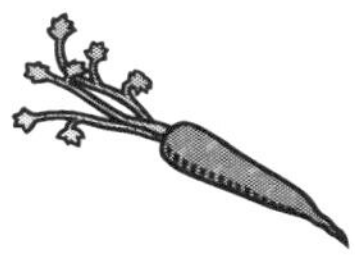	9	
	6	
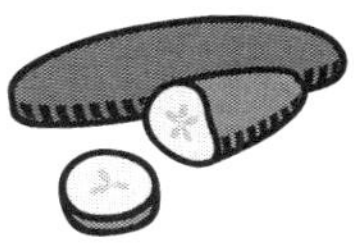	3	

2. Circle the most popular vegetable.

3. Circle the least popular vegetable.

BRAIN STRETCH

Make two correct number sentences using addition.

9 4 5

_______ + _______ = _______

_______ + _______ = _______

MONDAY Patterning and Algebra

1. Count on by 1s.

87, 88, 89, ________, ________, ________, ________

Use a number line or counters to help solve the following questions.

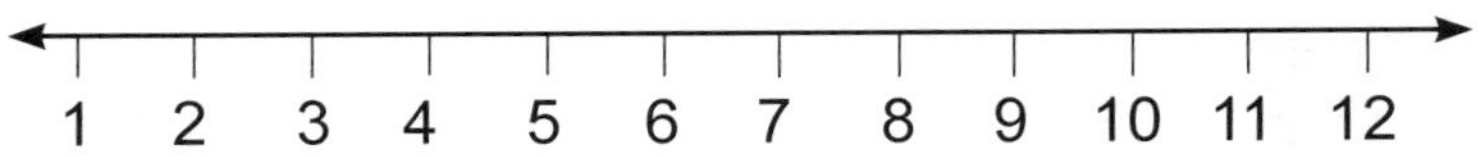

2. ________ + 6 = 7

3. 12 − ________ = 6

4. ________ + 5 = 9

5. 6 − 5 = ________

TUESDAY Number Sense and Operations

1. How many tens and ones?

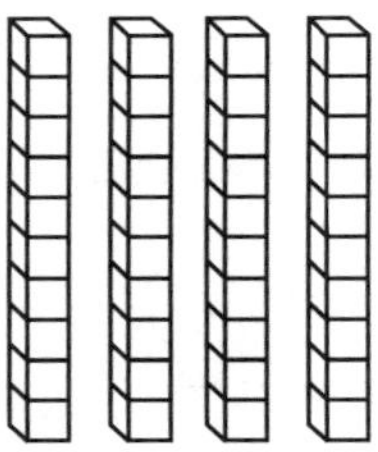

tens ______

ones ______

2. Write the numeral.

sixteen ______

3. Order the numbers from largest to smallest.

20, 36, 13

______, ______, ______

4. Find 10 more than 18. _____

Find 10 less than 18. _____

WEDNESDAY Geometry

1. Colour the cylinder orange. Colour the cube yellow.
   Colour the cone green. Colour the rectangular prism blue.

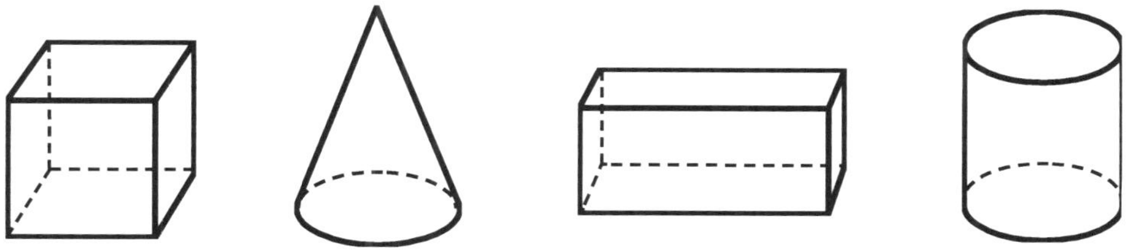

2. How many sides does this shape have?

_____

3. Draw a square and a rectangle. Tell one way they are the same.

____

Tell one way they are different.

____

THURSDAY Measurement

1. What time will it be 1 hour from now?

____ : ____

2. What month comes after February?

A. March B. December

3. Measure the length of the line.

It is about __________  long.

FRIDAY Data Management

1. Complete the tally chart.

Favourite Musical Activity

	Number	Tally
	9	
	6	
	3	

2. Circle the most popular activity.   

3. Circle the least popular activity.   

BRAIN STRETCH

Make two correct number sentences using subtraction.

6 2 8

________ − ________ = ________

________ − ________ = ________

Math — Show What You Know!

____

- [ ] I read the question and I know what I need to find.
- [ ] I drew a picture or a diagram to help solve the question.
- [ ] I showed all the steps in solving the question.
- [ ] I used math language to explain my thinking.

Student Tracking Sheet

Student	Week 1	Week 2	Week 3	Week 4	Week 5	Week 6	Week 7	Week 8	Week 9	Week 10	Week 11	Week 12	Week 13	Week 14	Week 15

Student Tracking Sheet

Student	Week 16	Week 17	Week 18	Week 19	Week 20	Week 21	Week 22	Week 23	Week 24	Week 25	Week 26	Week 27	Week 28	Week 29	Week 30

Math Star

Great Work!

Name ______________________________

Week 1, pages 1–3

Monday 1. △ 2. 5 in all

Tuesday 1. 1 ten, 3 ones 2. 4 3. 4 4. The first set should be circled.

Wednesday 1. circle 2. 1 3. 0

Thursday 1. 4 o'clock 2. The insulated jar should be circled. 3. A

Friday 1. 6 dogs, 2 birds, 5 cats 2. dog 3. bird

Brain Stretch 1. 3 + 6 = 9 2. 4 – 1 = 3

Week 2, pages 4–6

Monday 1. large triangle 2. 2 are left

Tuesday 1. 1 ten, 9 ones 2. 5¢ 3. 1 4. The first set should be circled.

Wednesday 1. triangle 2. 3 3. 3

Thursday 1. 12 o'clock 2. B 3. A

Friday 1. The Number column of the Favourite Fish chart should contain, from the top, 6, 10, 4.
2.  3. 

Brain Stretch 1. 4 + 4 = 8 2. 5 – 2 = 3

Week 3, pages 7–9

Monday 1. □ 2. 7 in all

Tuesday 1. 1 ten, 7 ones 2. 10¢ 3. 6 4. The first set should be circled.

Wednesday 1. square 2. 4 3. 4

Thursday 1. 6:00 2. A 3. 4

Friday 1. 2 tops, 4 Jack in the boxes, 5 bears 2.  3. 

Brain Stretch 1. 6 + 3 = 9 2. 6 – 3 = 3

Week 4, pages 10–12

Monday 1. △ 2. 9 in all

Tuesday 1. 2 tens, 1 one 2. 25¢ 3. 4 4. The second set should be circled.

Wednesday 1. rectangle 2. 4 3. 4

Thursday 1. 1:00 2. A 3. 7

Friday 1. 5, 7, 6 2.  3. 

Brain Stretch 1. 6 + 8 = 14 2. 8 – 2 = 6

Week 5, pages 13–15

Monday 1. Sample answer: (■□)■□■□ 2. 2 are left

Tuesday 1. 1 ten, 6 ones 2. 100¢ 3. 2 4. The first set should be circled.

Wednesday 1. pentagon 2. 5 3. 5

Thursday 1. 2 o'clock 2. C 3. 2

Friday 1. 5 strawberries, 3 watermelons, 6 bananas 2.  3. 

Brain Stretch A. One portion of the circle should be coloured. B. One portion of the rectangle should be coloured.

Week 6, pages 16–18

Monday	**1.** Accept any pattern. Sample answer: ▲△△▲△△▲ **2.** 3 are left
Tuesday	**1.** 3 tens, 4 ones **2.** 200¢ **3.** 10 **4.** 3
Wednesday	**1.** hexagon **2.** 6 **3.** 6
Thursday	**1.** 3:00 **2.** B **3.** 5
Friday	**1.** 6 carrots, 2 cauliflower, 8 cucumbers **2.** cucumber **3.** cauliflower
Brain Stretch	2, 3, 4, 5, 6, 7, 8, 9

Week 7, pages 19–21

Monday	**1.** Accept any pattern.
	2. Colours should match the equations written. Sample answer: 1 + 4 = 5, 2 + 3 = 5
Tuesday	**1.** 2 tens, 5 ones **2.** 10¢ **3.** 7 **4.** The first set should be circled.
Wednesday	**1.** octagon **2.** 8 **3.** 8
Thursday	**1.** 9 o'clock **2.** A **3.** 1
Friday	**1.** 4 slices of cake, 6 muffins, 5 cookies **2.**  **3.** 
Brain Stretch	**A.** 6 **B.** 8 **C.** 9

Week 8, pages 22–24

Monday	**1.** Accept any pattern.
	2. Colours should match the equations written. Sample answer: 3 + 3 = 6, 4 + 2 = 6
Tuesday	**1.** 3 tens, 1 one **2.** 15¢ **3.** 7 **4.** 3
Wednesday	**1.** trapezoid **2.** 4 **3.** 4
Thursday	**1.** 5:00 **2.** 7 **3.** 5
Friday	**1.** 9, 3, 6 **2.**  **3.** 
Brain Stretch	**A.** False **B.** True Picture should show 6 = 7 – 1.

Week 9, pages 25–27

Monday	**1.** Accept any pattern of alternating shapes.
	2. Colours and lines should match the equations written. Sample answer: 7 – 3 = 4, 7 – 1 = 6
Tuesday	**1.** 6 tens, 5 ones **2.** 10¢ **3.** 8 **4.** 12
Wednesday	Shapes should be coloured: yellow triangle, blue circle, red rectangle, green square
Thursday	**1.** 7:00 **2.** A **3.** 3
Friday	**1.** 4 sandals, 6 shoes, 3 boots **2.**  **3.** 
Brain Stretch	3 + 6 = 9; 5

Week 10, pages 28–30

Monday	**1.** Sample answer: □□○○□□
	2. Colours and lines should match the equations written. Sample answer: 8 – 5 = 3, 8 – 4 = 4
Tuesday	**1.** 4 tens, 2 ones **2.** 15¢ **3.** 9 **4.** 1 ten 
Wednesday	4 circles, 3 triangles, 5 squares
Thursday	**1.** 11 o'clock **2.** A **3.** 8
Friday	**1.** 6, 7, 9 **2.** 
Brain Stretch	**1.** 9 **2.** 3 **3.** 3 **4.** 12

Week 11, pages 31–33

Monday **1.** Look for blocks coloured to show 1 + 3 = 4, 3 + 1 = 4, and 2 + 2 = 4, possibly 0 + 4 = 4 or 4 + 0 = 4. **2.** 4

Tuesday **1.** 5 tens, 8 ones **2.** 3 cubes; 1 ten, 3 ones **3.** 13

Wednesday **1.** The square, hexagon, trapezoid, and rectangle should be coloured.
2. The square, rectangle, and trapezoid should be coloured.

Thursday **1.** half past 8 **2.** B **3.** 5

Friday **1.** 3 squares, 5 circles, 4 triangles **2.** circle

Brain Stretch **1.** 1 **2.** 6 **3.** 2 **4.** 6

Week 12, pages 34–36

Monday **1.** Sample answer:  **2.** Accept any pattern of squares and triangles.
3. Ensure description matches pattern.

Tuesday **1.** 2 tens, 9 ones **2.** Sample answer: 2 dimes or 4 nickels **3.**  **4.** 9

Wednesday **1.** Shapes should be coloured, in order: red, green, yellow, blue **2.** A **3.** A

Thursday **1.** half past 5 **2.** A **3.** 2

Friday **1.** 5 volleyball, 2 soccer, 3 football **2.** soccer

Brain Stretch **1.** 8 **2.** 14 **3.** 16 **4.** 9

Week 13, pages 37–39

Monday **1.** Sample answer:  **2.** Look for blocks coloured to show 6 + 3 = 9, 7 + 2 = 9 and so on.

Tuesday **1.** 8 tens, 2 ones **2.** Sample answer: 3 nickels **3.** A **4.** A

Wednesday **1.** Sample answer:  **2.** A **3.** B

Thursday **1.** half past 10 **2.** A **3.** 12

Friday **1.**  **2.** 

Brain Stretch **1.** 8 **2.** 9 **3.** 5 **4.** 9

Week 14, pages 40–42

Monday **1.** Sample answer:  **2.** 7 + 2 = 9

Tuesday **1.** 4 tens, 0 ones **2.** 20¢ **3.** 20 **4.** 12

Wednesday 2 parallelograms, 4 hexagons, 4 squares

Thursday **1.** half past 3 **2.** B **3.** 11

Friday **1.** 2 colouring, 5 painting, 4 video games **2.**  **3.** 

Brain Stretch **1.** 12 **2.** 10 **3.** 6 **4.** 4

Week 15, pages 43–45

Monday **1.**  **2.** Look for blocks coloured to show 6 – 4 = 2, or 6 – 5 = 1 and so on.

Tuesday **1.** 3 tens, 3 ones **2.** 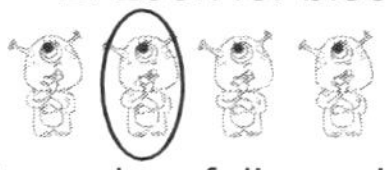 **3.** 17 **4.** 18 < 22

Wednesday Shapes should be coloured as follows: blue rectangle, green triangle, red trapezoid, and yellow square.

Thursday **1.** half past 1 **2.** A **3.** 7

Friday **1.** 4 trumpets, 2 baton twirlers, 6 drummers **2.**  **3.** 

Brain Stretch **1.** 11 **2.** 10 **3.** 8 **4.** 3

Week 16, pages 46–48

Monday 1. Sample answer:  2. 4 + 4 = 8 3. 2 + 2 = 4 4. 5 + 5 = 10

Tuesday 1. 9 tens, 0 ones 2. 17 3.  4. One part of the triangle should be coloured.

Wednesday 1. The 3 large triangles should be coloured. 2. A 3. A

Thursday 1. half past 8 2. B 3. 4

Friday 1. Graph should be shaded to 3 for the top fish, 2 for the middle fish and 5 for the bottom fish.
2.  3. 5 – 2 = 3

Brain Stretch 4, 6, 8, 10, 12

Week 17, pages 49–51

Monday 1. 9 2. 11

Tuesday 1. 6 tens, 9 ones 2. 20¢ 3. 11, 13, 15 4. 

Wednesday 1. The 3 large circles should be coloured. 2. B 3. B

Thursday 1. half past 12 2. A 3. 7

Friday 1. Shading should extend to 3 for dogs, 2 for birds and 5 for cats. 2. 3 dogs, 2 birds and 5 cats
3. 5 – 3 = 2

Brain Stretch 10, 15, 20, 25, 30

Week 18, pages 52–54

Monday 1. 13 2. 14

Tuesday 1. 7 tens, 5 ones 2. 40¢ 3. 18, 13, 11 4. One part of the shape should be coloured.

Wednesday 1. Shapes should be coloured, in order: blue, red, blue, yellow, green, red, yellow 2. A 3. a triangle

Thursday 1. half past 2 2. A 3. 3

Friday 1. Graph should be shaded to 7 for cake, 5 for cupcake and 8 for cookie
2. 7 cake, 5 cupcake, 8 cookie 3. 7 – 5 = 2

Brain Stretch 1 group of ten, 5 left over, 15

Week 19, pages 55–57

Monday 1. Patterns will vary. 2. 7 3. 8 4. 10 5. 1

Tuesday 1. 3 tens, 2 ones 2. 19 3. 20, 14, 12 4. One section of the shape should be coloured.

Wednesday 1. Shapes should be coloured, in order: red rectangle, yellow pentagon, green square, red trapezoid, yellow pentagon, red rectangle, red trapezoid. 2. A 3. a square

Thursday 1. half past 7 2. B 3. 6

Friday 1. Graph should be shaded to 4 for hexagon, 2 for heart and 5 for circle. 2. 11

Brain Stretch 4 + 3 + 6 = 13; 4 + 6 + 3 = 13 Sample answer:

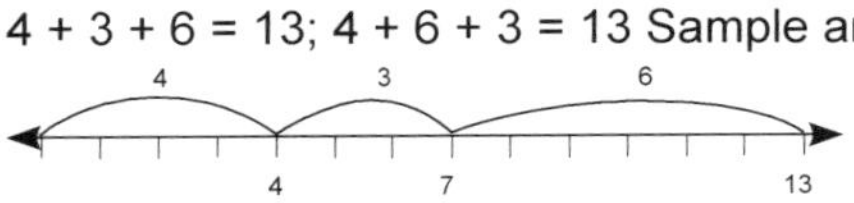

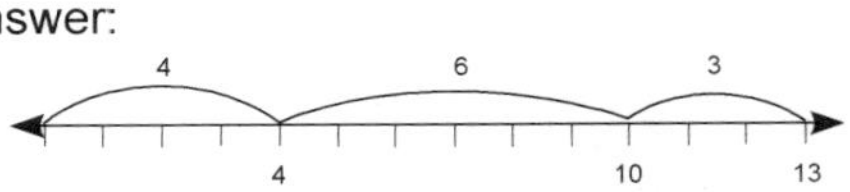

Week 20, pages 58–60

Monday 1. 15 2. 17 3. 9 4. 6

Tuesday 1. 9 tens, 5 ones 2. 11 3. 30, 26, 11 4. 35 + 10 = 45 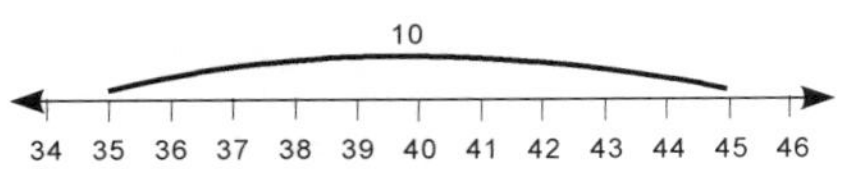

Wednesday 1. Shapes should be coloured, in order: red, blue, green, orange, green, red, orange 2. A
3. a rectangle

Thursday 1. half past 9 2. A 3. 7

Friday 1. Bar graph shading should extend to 4 for lollipops, 3 for ice cream and 4 for cotton candy
2. 4 lollipops, 3 ice cream, 4 cotton candy

Brain Stretch 1 group of ten, 6 left over, number 16

Week 21, pages 61–63

Monday **1.** Patterns will vary. **2.** 3 + 7 + 8 = 18; 3 + 7 = 10; 18 cards

Tuesday **1.** 6 tens, 6 ones **2.** 15¢ **3.** 3 **4.** one half; two halves

Wednesday **1.** The pentagon and octagon should be coloured. **2.** B **3.** A

Thursday **1.** half past 11 **2.** 12 **3.** 5

Friday **1.** Dog 𝍸, Bird |||, Cat 𝍸 || **2.**  **3.** 

Brain Stretch 15; 15 bars

Week 22, pages 64–66

Monday **1.** Sample answer:  **2.** 9 **3.** 1 **4.** 12 **5.** 0

Tuesday **1.** 6 tens, 4 ones **2.** 15¢ **3.** 0 **4.** One third of the shape should be coloured.

Wednesday **1.** The rectangle, square and parallelogram should be coloured. **2.** A **3.** A

Thursday **1.** half past 7 **2.** A **3.** 6

Friday **1.** top row 𝍸 |, middle row ||||, bottom row 𝍸 **2.**  **3.** 

Brain Stretch 4 – 1 = 3

Week 23, pages 67–69

Monday **1.** Patterns will vary. **2.** 2 **3.** 1 **4.** 9 **5.** 6

Tuesday **1.** 8 tens, 4 ones **2.** 60 **3.** 7 **4.** one fourth; four fourths

Wednesday **1.** The pentagon, rectangle and triangle should be coloured. **2.** B **3.** B

Thursday **1.** half past 12 **2.** The bars should be coloured in the following order: yellow, blue, red. **3.** 5

Friday **1.** Slide ||, Teeter totter 𝍸 |||, Carousel |||| **2.**  **3.** 

Brain Stretch 5 + 3 = 8; 3 more

Week 24, pages 70–72

Monday **1.** Patterns will vary. **2.** 12 **3.** 2 **4.** 8 **5.** 4

Tuesday **1.** 4 tens, 0 ones **2.** 7 **3.** 20 **4.** 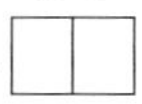 one half; two halves

Wednesday **1.** The pentagon, hexagon and octagon should be coloured. **2.** A **3.** B

Thursday **1.** 9:00 **2.** A **3.** B

Friday **1.** Square ||, Circle 𝍸 ||, Triangle 𝍸 | **2.** circle **3.** square

Brain Stretch 10 – 7 = 3; 3 fewer pens

Week 25, pages 73–75

Monday **1.** 62, 61, 60, 59 **2.** 13 **3.** 1 **4.** 10 **5.** 9

Tuesday **1.** 7 tens, 8 ones **2.** 12 **3.** 5 **4.** 1 tens, 8 ones = 18

Wednesday **1.** The circle, triangle and square should be coloured. **2.** A **3.** A

Thursday **1.** 12:30 **2.** A **3.** A

Friday **1.** Colouring 𝍸 |||, Painting 𝍸 |, Video games || **2.**  **3.** 

Brain Stretch 5 + 2 = 7

Week 26, pages 76–78

Monday 1. 28, 30, 32, 34 2. 8 3. 8 4. 9 5. 1
Tuesday 1. 2 tens, 5 ones 2. 7 3. 14 4. A
Wednesday 1. A 2.  3. B 4. B
Thursday 1. 9:30 2. B 3. B
Friday 1. Snowman building 𝍸, Snow angel making |||, Tobogganing 𝍸 |||| 2.  3. 
Brain Stretch 1. True 2. True

Week 27, pages 79–81

Monday 1. 45, 50, 55, 60 2. 1 3. 9 4. 7 5. 5
Tuesday 1. 3 tens, 3 ones 2. 50 3. 6 4. 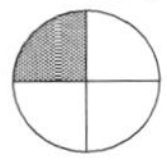
Wednesday 1. B 2. B 3. A
Thursday 1. 7:00 2. A 3. A
Friday 1. Strawberries ||||, Watermelon 𝍸 |||, Bananas 𝍸 2. watermelon 3. strawberries
Brain Stretch 1. True 2. False

Week 28, pages 82–84

Monday 1. 69, 70, 71, 72 2. 7 3. 5 4. 10 5. 5
Tuesday 1. 4 tens, 4 ones 2. 72 3. 9 4. one fourth
Wednesday 1. A 2. A 3. Sample answer: It has 4 sides. 4. 0
Thursday 1. 11:00 2. B 3. No, A and B are the same length.
Friday 1. Top ||, Jack in the box 𝍸 |, Teddy bear 𝍸 |||, 2. teddy bear 3. spinning top
Brain Stretch 7 – 4 = 3, 7 – 3 = 4

Week 29, pages 85–87

Monday 1. 70, 80, 90, 100 2. 5 3. 4 4. 5 5. 8
Tuesday 1. 1 ten, 6 ones 2. 25¢ 3. 33, 23, 3 4. B
Wednesday 1. The rectangular prism and the pyramid should be red, the sphere should be blue and the cylinder should be green.
Thursday 1. 2:30 2. A 3. 5
Friday 1. Carrot 𝍸 ||||, Cauliflower 𝍸 |, Cucumber ||| 2. carrot 3. cucumber
Brain Stretch 5 + 4 = 9, 4 + 5 = 9

Week 30, pages 88–90

Monday 1. 90, 91, 92, 93 2. 1 3. 6 4. 4 5. 1
Tuesday 1. 4 tens, 0 ones 2. 16 3. 36, 20, 13 4. 28; 8
Wednesday 1. The shapes should be coloured, in order: yellow, green, blue, orange 2. 3
3. 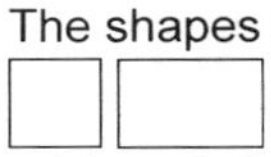 Sample answer: Both shapes have 4 sides. The square has 4 sides equal in length.
Thursday 1. 4:00 2. A 3. 7
Friday 1. Trumpet playing 𝍸 ||||, Marching band 𝍸 |, Drumming ||| 2. playing trumpet 3. drumming
Brain Stretch 8 – 2 = 6, 8 – 6 = 2